CONTENTS

Preface vii

Introduction 1
    Carolyn W. Pumphrey

1. Energy and Security Keynote 23
    Alan Hegburg
    — Questions and Answers 34

2. Fossil Fuels 51
    — Oil and Global Security 51
        Anne Korin
    — Protecting the Prize 63
        Eugene Gholz
    — Coal, Climate Change, and Conflict 73
        Kevin Book
    — Commentary 83
        Rosemary Kelanic
    — Questions and Answers 91

3. Alternative Energy: Nuclear and Water 107
    — Introduction 107
        Alex Roland
    — Nuclear Proliferation 108
        Steven N. Miller
    — Pathways between Civilian and
        Military Nuclear Power 116
        Man-Sung Yim
    — The Water-Energy-Security Nexus 121
        Carey King
    — Nuclear Energy and the Military 130
        James Bartis

— Energy, Environment, and Security 134
   Alex Roland, James Bartis,
   Carey King, Steven Miller,
   and Man-Sung Yim
— Comparative Security of Different
   Technologies 138
   Alex Roland, James Bartis,
   Carey King, Steven Miller,
   and Man-Sung Yim
— Questions and Answers 142

4. Alternative Energy from a Security
Perspective 157
   — Hidden Costs of Energy 157
      James Trainham, James Bartis,
      David Dayton, Michael Roberts,
      and Daniel Weiss
   — Climate vs. Economics: Security
      Implications of Energy Choices 166
      James Trainham, James Bartis,
      David Dayton, Michael Roberts,
      and Daniel Weiss
   — Shale Gas: A Key to Energy Security?.....177
      James Trainham, James Bartis,
      David Dayton, Michael Roberts,
      and Daniel Weiss
   — Alternative Energy and International
      Relations 182
      James Trainham, James Bartis,
      David Dayton, Michael Roberts,
      and Daniel Weiss
   — Questions and Answers 188

5. The Political Environment and
   U.S. Energy Security 203
   — Rising Great Powers and Competition
     for Energy 203
     Bernard Cole
   — Unconventional Threats to Energy
     Supplies 216
     Robert F. Cekuta
   — Smart Grid Vulnerabilities to
     Cyber Attacks 224
     John Bumgarner
   — Commentary 232
     Stephen Kelly
   — Questions and Answers 235

6. Solutions 253
   *Vikram Rao, William Boettcher, and
   Douglas Lovelace*

About the Contributors 291

PREFACE

During the period March 3-4, 2011, the Triangle Institute for Security Studies (TISS), North Carolina State University (NCSU), and the Strategic Studies Institute (SSI) held a colloquium at the McKimmon Center in Raleigh. The event received additional financial support from the Research Triangle Energy Consortium, the Oak Ridge National Laboratory-North Carolina State University Collaborative Research Program, and Duke University's Program in American Grand Strategy. It was launched as part of the TISS/NCSU Energy and Security Initiative which was formed in 2010 and is dedicated to exploring the links between energy and security.

The colloquium, entitled "The Energy and Security Nexus: A Strategic Dilemma," was attended by 128 persons from federal and state government, academia, think tanks, and a wide variety of local organizations and businesses working on energy issues. The goal of this conference was to explore the connections between energy and security (human, national, and international) and to consider how best to resolve strategic dilemmas.

This edited volume is based on this event. It reflects, as closely as possible, the form and content of the colloquium. However, the presentations have been adapted to be read. In the case of the discussion, some portions were not adequately captured on tape and so were omitted, while elsewhere the editor somewhat modified the text in the interest of clarity and brevity. Notwithstanding, the proceedings remain true to the spirit and form of the conference.

The relationship between energy and security has been receiving increasing attention over the last few years, and the colloquium provided some timely insights. TISS would like to thank all those who made it possible. Too many individuals and institutions provided conceptual and other help along the way to be acknowledged here. But we would like to acknowledge a few in particular. First, we owe thanks to the U.S. Army War College (USAWC) for its generous support of this project. Second, we owe thanks to the participants. Their collegiality and professionalism made the conference a success.

Besides these, the editor would like to express her personal gratitude to a number of individuals. Raymond Fornes and Christopher Gould (The School of Physical and Mathematical Sciences, North Carolina State University) helped frame the conference and provided financial support. Without their encouragement, this project would never have been brought to its conclusion. Richard Kearney (School of Public and International Affairs), Lorenzo Wilson (Department of Horticulture), Man-Sung Yim, (Department of Nuclear Engineering), and William Boettcher (Department of Political Science) were key faculty members in the Energy and Security Initiative at North Carolina State University, on whose excellent advice I relied. Vikram Rao, Research Triangle Energy Consortium, provided creative insights into the conference design and brought to the table the many interesting technical experts who enriched our discussion. Ambassador Patrick Duddy, though unfortunately unable to serve as a participant, was tireless in his efforts on my behalf and helped secure some of the key speakers. Joseph Caddell (University of North Carolina at Chapel Hill), Joseph Caddell, Jr., and Alex Roland (Duke University) read portions of this manuscript and provided constructive criticism where it was most needed.

Joseph Caddell also provided crucial guidance in framing the conference proposal. Finally, my thanks go to faculty members at SSI. Douglas Lovelace was a much valued participant on the final panel, and Dallas Owens, as always, was the best possible of colleagues, offering invaluable advice and patiently answering all my questions.

      Carolyn W. Pumphrey

INTRODUCTION

Carolyn W. Pumphrey

It is hard to overstate the importance of energy.[1] Energy literally drives the global economy. Societies rely on it for everything from advanced medical equipment to heating, cooling, and irrigation. Whether it derives from advanced nuclear reactors in developed nations or simple woodstoves in the developing world, energy is recognized as vital to human welfare. It influences our economic, political, and social policies. Possessing or not possessing sufficient energy is a key determinant of a state's political and economic power. Competition for energy has been, is, and will be a source of conflict. And the choices we make when it comes to energy will have a profound bearing on a wide range of security concerns, from nuclear proliferation to climate change.

The joint Triangle Institute for Security Studies (TISS)/North Carolina State University (NCSU)/Strategic Studies Institute (SSI) conference on the "Energy and Security Nexus," which was held on March 3-4, 2011, addressed many of these issues. To place the remarks made at this conference within a clearer context, a few terms and concepts need to be discussed. To provide a further framework for the reader, these definitions will be followed by a brief threat assessment, a cost-benefit analysis of different energy technologies seen from a security perspective, and an overview of the chief findings of the conference.[2]

DEFINITIONS

Security.

Security is traditionally divided into the international, national, and personal or human levels.[3] All relate to the "avoidance of harm." Human security is the protection of the individual from harm. National security deals with the need of a nation-state to protect itself from harm, whether internal or external. International security usually addresses the problems associated with groups of nation-states working to ensure international stability and the avoidance of dysfunctional events or conflicts that will affect nations and individuals. All levels ultimately relate back to the concept of human or personal security. In the case of energy security, the international, national, and personal energy needs are affected by international, national, and individual actions. In human existence, these levels of security cannot be separated.

Energy Security.

Energy security means rather different things to different people. However, for working purposes, it may be said to include three components: reliability, affordability, and environmental friendliness.[4] Reliability means that a state has regular, noninterrupted access to energy in the amount and shape it needs. Affordability means that it has access to energy supplies at a price that can be sustained economically and promotes economic growth.[5] Environmental friendliness means that a state relies primarily on energy that provides for environmental sustainability and limits destructive social results. It is now also equated by many with the use of forms of energy that do not release climate-changing gasses.[6]

EVALUATING THE THREAT

At present, the United States enjoys a relatively high level of energy security, at least in the sense that our energy is affordable and reliable. Prices at the pump have been high of late, and this does hurt the average citizen, especially in a time of economic recession. But there is no cause for alarm.[7]

The stability and reliability of our supplies stem from increased efficiency, new finds of natural gas, and better mechanisms for dealing with problems than we had in the past. These include the existence of a Strategic Petroleum Reserve and the creation in 1994 of the International Energy Agency, an institution which makes possible timely collective responses to politically inspired disruptions.[8] The location of key energy supplies in politically unstable regions deters investment and creates some constraints. However, contrary to widespread popular perception, our supplies are not significantly threatened by terrorism or international crime.[9] The market, moreover, generally compensates for short-term interruptions in supply.[10]

We do have some security vulnerabilities, however. In the first place, many other nations lack crucial natural resources or are faced with rapid population expansion. Their drive to improve their energy security has the potential to undermine U.S. security. Instability can foster terrorism and international crime and even lead to militant strategies. The need to secure badly needed energy supplies may also lead some nations to take steps that harm collective interests. They may, for example, rely on cheap and unclean sources of energy like coal, thereby aggravating climate-change problems.[11]

The United States also pays a price for its heavy dependence on foreign oil in its transportation sector. Reliance on foreign oil means that money and jobs flow out of the United States. Petrodollars fuel corruption, hamstring our foreign policy, and fund institutions fostering radical ideologies.[12] We are less vulnerable than formerly to the use of energy as a weapon because multinational corporations are more driven by market than political considerations and because countries are rarely reliant on just one state for supplies or for access to supplies. But it can still be a dangerous tool in the hands of our enemies.

A new challenge stems from our increasing dependence on centralized electric grids.[13] For electricity, we depend on things like hydropower, coal-fired and gas-fired generators, nuclear power plants, and even wind mills. All of these can be physically broken by computer manipulation. Our way of life could be seriously undermined by failure of these grids, whether caused by accident or design.

The international community as a whole has in general met with limited success in meeting the third component of energy security: environmental friendliness. From the radioactive waste of nuclear power plants to the forests cut down to produce wood fuel, the production and use of energy cause ecological damage. Energy use has also led to the increased input of carbon dioxide into the atmosphere and hastened climate change.[14] This in turn has the potential to create a variety of security problems. Such is particularly true in the case of vulnerable and underprivileged parts of the world where societies have limited ability to mitigate climate change impact.[15]

The Future.

Future energy security challenges look more daunting. This is less because of shrinking supplies than it is because of growing demand. Analysts disagree over crucial factors such as whether we have reached peak oil production or whether we will find technological solutions to our problems.[16] They agree that the human population is rising sharply as are hopes for better standards of living among the underprivileged. These factors in combination are likely to increase competition for energy and potentially create some significant security problems.[17]

TECHNICAL OPTIONS

Energy is inextricably tied to technology. However, the technologies currently at our disposal often have major security costs.

Oil.

This queen of fuels is of fundamental importance in the U.S. transportation sector. It is energy-rich, cost-effective, and supported by an existing infrastructure.[18] It is vital for the Armed Forces which put a premium on military effectiveness.[19] It at least appears to be cheap, though some stress that the price we pay at the pump does not take into consideration the actual cost of securing and cleaning up after it, let alone fighting wars over it.[20]

Stacked against these benefits are the security problems oil creates. Where extraction occurs in thoughtless fashion or inadequate security precautions are taken, production can wreak havoc on local communities. Transportation through pipelines is also feared by many because of contamination/pollution potential—hence the opposition in the United States to the proposed Keystone pipeline.[21] Most significantly, by releasing carbons, the burning of oil hastens global warming. Of all the security arguments against oil, this may be the most compelling.

Coal.

In the United States, coal currently generates about half of our electricity and about 21 percent of all our energy needs. Although, like all fossil fuels, its supply is finite, it is for now relatively abundant. It is easy to process which makes it especially valuable in parts of the world where technological capabilities are limited.[22] It is also cheap.

At the same time, it is a threat to human, national, and collective security. Depending on mining techniques used, it can be devastating to the environment. It is dangerous to extract and dangerous to health. It yields more carbon dioxide than even petroleum.

Clean coal may help solve our problems in the future. But as of this time, the integrated gasification combined cycle (IGCC) and other methods are too expensive to be truly viable. Coal-to-liquid technology is environmentally messy, and carbon sequestration may pollute ground water.

Gas.

Natural gas promises to help solve some of our security problems.[23] Its energy content is high, if not as high as oil. We have substantial reserves of shale gas, especially along the east coast and in Texas. Fostering a domestic industry would reduce our need to import both oil and the potentially deadly liquefied natural gas (LNG).[24]

Unfortunately, getting natural gas out of the ground and getting it where it needs to go are riddled with controversy. Opponents of hydraulic fracturing claim that it causes all kinds of damage, including contamination of water supplies (in some areas) and diversion of precious water resources (in others). Hydraulic fracturing also releases fugitive methane. So despite the fact that natural gas emits relatively few carbons when burned, it is not the most obvious solution to global warming.[25]

Wind and Sun.

Wind and sun have some security benefits. They are clean and renewable and can reduce dependence on centralized grids. Military bases that have their own supply of energy from sun and wind have something to turn to in the case of electrical power outages.[26] The same goes for the civilian sector. If we are worried about cyber attacks on the electric grid, having homes equipped with their own solar panels makes a lot of sense.[27]

However, both wind and sun are intermittent and have to be supplemented by more reliable sources of power. Some places have more sun and wind than others. Wind mills and solar panels are not cheap, in part because our current infrastructure is geared toward fossil fuels. For all these reasons, there seems to be little prospect of their being able to replace fossil fuels in the near future.

Biomass.

Biomass is a promising form of alternative energy, which might be able to meet 10-15 percent of U.S. energy needs fairly soon. Its high cost is a function of factors that might change: lack of existing infrastructure, hidden subsidies of fossil fuels which harm competition, and a small market. Biomass has rather low energy density and so is of limited use to the military. But there are other more significant security concerns. Fuels like ethanol are derived from crops. To grow them, land must be cleared, which often entails deforestation, and the fields must be fertilized. All these things result in the emission of greenhouse gasses. Moreover, the production of biofuels uses 20 to 30 times as much water as the production of gasoline.[28] Where water is plentiful, this may not be a game changer, but if world supplies of water diminish in the future, this might be a source of major conflict.

Critics of biofuels further claim that diversion of land for fuel in the United States has driven up world food prices and arguably triggered uprisings in the Middle East. Though their arguments are contested, they should not be discounted without further research.[29]

Happily, there are other kinds of biomass which do not create the same set of security problems. Switchgrass, for example, is grown on land not suited for crops and does not require fertilizer or irrigation. Algae produce energy through the process of photosynthesis. Though they give off carbon dioxide (CO2) when burned, they absorb CO2 while they are growing, and they do not require much land.[30]

Nuclear.

The economics of nuclear power are hotly debated and complicated by the existence of a variety of different processes, some of which, like reprocessing, might change the commercial calculation for the better. Nuclear power plants take a long time to build but once built, they are a rather long-lived asset. Nuclear power has undeniable advantages. It is a steady source of power and is capable of supplying a sizable portion of our electric needs.[31] The U.S. military is also looking into the possible use of small modular reactors which could be assembled on base and may even be used for forward deployment.[32] Far the most important security benefit, however, is the fact that nuclear energy does not produce greenhouse gasses.[33]

Nuclear power is, however, beset by problems. Despite a good security record, nuclear accidents have happened, and have made nuclear power unpopular among many communities.[34] Dealing with nuclear waste is another difficult issue. While the technical community thinks it has the means to solve the problems, others are skeptical. In the United States, the Yucca Mountain option was set aside because of both engineering and political problems. Breeder reactors which theoretically recycle and consume all actinides may provide a solution. But they arguably carry a proliferation risk. Steps can be taken to safeguard and secure facilities. However, humans do fail, and companies are frequently hesitant to incur the expense of making facilities ultra secure.[35]

These problems pale in comparison with some others. What would happen, for example, were conflicts to arise in a region with a lot of nuclear assets? What if nuclear facilities are targeted — as indeed was done by Iran during the Iran-Iraq war? Finally, there are unanswered questions about the pathway between civilian nuclear power and nuclear weapons development. Some think that countries eager to develop a successful nuclear industry will avoid taking any actions that might lead to the imposition of sanctions or worse. Others, however, think that the proliferation risk is real.[36]

Water.

The United States generates about 6 percent of its electricity from water in the form of hydroelectric energy, but water also plays an important role in the production and distribution of other forms of energy. Hydroelectric power has all the advantages of other alternatives including that of being clean. By the same token, it can cause extensive environmental problems, especially if not conducted in a sustainable way. Dams can be disrupted by cyber attacks, and water can shrink in times of drought.[37]

Worst of all, there are many other water needs. Water is needed to cool power plants, to irrigate biofuels, to extract fossil fuels from the ground—and the list goes on. Moreover, energy is needed for a variety of crucial functions from making water drinkable to getting it where is most needed. Both water and energy are incredibly important for human use, and both are limited resources. We may have to make some very hard choices down the road.[38]

THE STRATEGIC DILEMMA

The strategic dilemma that faces us is visible in the first National Security Strategy of the Barack Obama Administration, released in May 2010. The authors of this document see the transformation of the way we use energy as a key to our economic revitalization (and ultimately our safety) as well as a way to mitigate the security problems many expect to follow from climate change. The 2010 Strategy recognizes the danger of disruption to our energy supplies and the political vulnerability that can accompany dependency. At the same time, the strategy stresses the need to reduce the budget deficit. It sees the promotion of international human rights and food security as a key to international order and names nuclear proliferation as the single most dangerous threat to the American people. It insists that the United States must maintain its conventional superiority, enhance its ability to defeat asymmetric threats, and preserve its nuclear deterrent capability as long as other nations possess nuclear weapons.[39] However, these goals can work at cross purposes. In implementing any one of these strategic goals, we run the risk of undermining another. Depending upon our priorities, we are likely to push for different kinds of energy technologies, and every step of the way our decisions will be made the harder because of the many still unanswered questions about the technologies themselves.

If we are sure that our foreign policy is being driven by excessive dependence on fossil fuels, we will do all that we can to foster alternatives. We will also favor investment in alternatives if we see climate change as *the* threat of the future. But if we do this, we are likely to run up big bills and perhaps find ourselves with a less effective military. If we choose the wrong kind of alternatives, we may drive up world food prices, encourage deforestation, and ironically lead the world down the very path of climate-driven instability we are trying to avoid.

If we see the root of all our power as economic and wish to avoid excessive debt at all costs, we will be reluctant to wean ourselves off oil. But down the road, we will have to pay for the security costs of climate change. We are also likely to have less freedom of action in our foreign policy, and we may also be providing financial aid to terrorists and other foreign enemies.

If we view nuclear proliferation as *the* most pressing security concern of our day, we may hesitate to promote civilian nuclear industry, and yet it is just possible that this source might solve both our need for clean energy and reduce proliferation.

It is indeed a dilemma.

SOLUTIONS

A number of strategies to improve energy security were discussed at the conference. Among them were the use of force, cooperation, and the expansion of supply through the means of technology. What conclusions did participants reach?

The Use of Force.

The use of military force to protect our energy security is controversial. American military power provides the stability that allows the global oil market to function.[40] At the same time, there are clear limits to what it can accomplish. It cannot affect growing energy demand. Nor can it prevent "peak oil." Some scholars further argue that forward military deployment is not the best answer for politically-motivated disruptions of supply. Radical disruptions might result from the conquest of the Middle East, but this is an unlikely scenario. They might also result from civil wars within an oil-rich nation—but intervention in this case would be counterproductive. Most disruptions are likely to be temporary because market forces naturally compensate for obstructions in the flow of oil. If this logic is correct, this is good news, because a strategy of restraint will help us in a time of fiscal retrenchment.[41]

Cooperation and Reform.

One strategy that has already proved useful in improving energy security is to advance economic interdependencies. When consumer nations get their resources from multiple sources, supplier nations have less ability to use their assets as weapons. This should be encouraged. So too should cooperation. For example, states on the Persian Gulf faced by limited water supplies might benefit from sharing power plants. The United States would be well advised to work with the Mexicans to share fields that straddle their borders.[42]

Another strategy is to create the kind of environment that will promote a healthier energy industry. Supplier states like Iraq need to enact laws that will persuade investors that money flowing into the country will not fuel corruption.[43] States like Nigeria need to give their people a stake in industrial development to reduce unrest and increase productivity. The propagation of good business practices as well as state of the art technologies should be encouraged. In the oil business, for example, knowledge about good field management and engineering can be shared, to the benefit of all concerned.[44] Sharing advanced, safe, and "green" technologies with developing nations promises to pay security dividends.[45]

Expansion of Supply.

The third strategy is to enlarge the energy pie by creating new (alternative) forms of energy, developing new and better ways to exploit older ones, and by promoting efficiency and conservation. There are compelling reasons to develop alternatives as rapidly as we can. First and foremost, alternative energy is key to dealing with the climate-change problem. Second, it will help to ensure that more and cheaper fossil fuels can be reserved for use by the military in wartime. Third, the spread of alternative energy might allow developing nations to break into new markets.[46] Finally, reducing dependence on fossil fuels in the transportation sector will help liberate our foreign policy. These arguments are especially applicable to renewables. Because of the risks outlined above, the extent to which nuclear energy should be part of the picture remains debated.

Because the United States has extensive domestic natural gas reserves and because of the security advantages to be derived from reduced dependency on oil, shale gas has its strong supporters. There is also considerable support for the development of new technologies for cleaning coal which, though not a renewable fuel, is abundant. Opponents, however, resist continued reliance on fossil fuels and worry about the environmental damage done by extraction and disposal.

Conservation and efficiency are a vital part of any strategy that focuses on the expansion of supply through technology. Greater efficiency can be achieved by a wide variety of means. Many of these carry little risk—like the use of more energy-efficient building materials in our homes. Others, like recycling waste in breeder reactors, are more problematic.

MAKING THE STRATEGY WORK

The expansion of supplies through technology was the strategy that had the most resonance at this particular conference. It clearly has much to recommend it. Equally clearly, it will fail if two conditions are not met. First, technological development must go hand in hand with a stringent concern for safety and security. Second, we must find ways to change dysfunctional patterns of behavior.

Safety and Security.

This is especially critical in the case of nuclear power which promises so much but has such serious security implications. Means vary, but the case of Abu Dhabi shows us what can be done. This country, the first in the Middle East to integrate a nuclear power plant into the grid, negotiated a 1-2-3 agreement with the United States: It forsook recycling, agreed not to enrich, and adhered to safeguards.[47]

The same rigor must be applied in the case of other technologies. We must ensure that we maintain vigilant oversight and establish standards for protecting ground water and surface water, reducing air pollution, and capturing methane.[48] We must not allow our eagerness to produce energy at home and to make big profits blind us to the long-term costs of unwise development.

To ensure that we do not create one problem while attempting to solve another, we must improve our understanding of the security ramifications of specific technologies. Does nuclear recycling alleviate or add to the dangers of nuclear proliferation? What really is causing price rises and riots: biofuels production or speculation? Is hydraulic fracturing a real threat? These are but a few of the more obvious controversies.

We also need to better understand the political, cultural, and social forces that affect our quest for energy security. Questions abound: Will the growth of civilian nuclear power inhibit or expand the proliferation of nuclear weapons? What is the nature of the resource curse or the relationship between disorder and poverty? How can we measure risk? How much risk is acceptable?

Further research will be costly, to be sure, and there is a natural reluctance to spend money on research during a time of fiscal constraint. But it is costly, too, to move ahead without adequate knowledge of long-term security risks.[49]

Changing Patterns of Behavior.

Developing useful new technologies is one thing. Getting them adopted is another. It is a question of will, and the will is not always there.

Part of the answer, though not one everyone is happy about, is regulation. Some point out that the government is less efficient than the market in picking technologies. Others think that regulation is vital. Companies are by definition geared to profit and U.S. citizens are—so some would say—by culture a frontier people all too comfortable with waste.[50] Regulation may be part of the answer.

Another option is to play to the economic animal in humans by making alternatives more economically attractive. This can be done by subsidies, or by factoring into the cost of fuels those hidden costs of use and development. These include what we pay to repair damage to health and environment, to create the needed infrastructure, and to defend our supplies.[51] That would level the playing field for alternatives and help create new markets.

Yet again, we could model ourselves on the Brazilians, and become a nation of flex cars. With minor modifications, cars could be made to run on more than one type of fuel. That might give drivers the ability to choose the most economic fuel available at any point in time, break the stranglehold of oil on our transportation sector, and/or stimulate the development of alternative energy.[52]

Finally, there is a need for further education. This is going to be needed no matter what course we choose, particularly a public education campaign. We need to promote a better understanding of the concept of risk and a greater awareness of the fact that, when it comes to energy, there is no universal panacea, only trade-offs. This is vital if we are to create a constituency for more effective energy policies.

CONCLUSION

Our energy problems are challenging, but they are not insurmountable. The technologies involved are improving, and many of the associated risks can be limited if we move forward with caution and with an eye to possible security repercussions. As a nation, we are fortunate and can afford to engage in the kind of collaborative efforts that will help solve the world's collective security problems. At the same time, the stakes are high. If we make unwise decisions, the costs could be enormous, especially if we are talking about nuclear conflict or catastrophic climate change. We ignore these threats at our peril.

ENDNOTES - INTRODUCTION

1. From the Greek word "energeia," meaning "activity, operation," energy is an indirectly observed quantity. It is often understood as the ability of a physical system to do work on other physical systems.

2. The views offered in this introduction reflect those of participants in this conference. The reader interested in putting these views in a broader perspective may turn to the reference handbook, by Gal Luft and Anne Korin, eds., *Energy Security Challenges for the 21st Century*, Santa Barbara, CA: ABC Clio, 2009.

3. Some scholars prefer to use the term "collective" to stress the existence of problems that transcend national boundaries and cannot be solved without collective action. Among these is Gayle Smith, Special Assistant to the President and Senior Director of the National Security Council.

4. Brenda Shaffer, *Energy Politics,* Philadelphia, PA: University of Pennsylvania Press, 2009, p. 93.

5. Some scholars think high prices undermine energy security. Others focus on "price stability." They argue that we can adapt to "expensive" supply costs if we avoid price shocks and excessive volatility. (See, for example, the different views of Anne Korin and Eugene Gholz, Panel I, chap. 2.)

6. Shaffer, p. 93.

7. This reflects the consensus view of persons at the conference but was especially prominent in the arguments of Eugene Gholz (chap. 2) and Alan Hegburg (chap. 1).

8. See especially *Ibid.*

9. There are scholars and analysts who are quite concerned about the dangers posed by terrorism to energy security, but Robert Cekuta and Bernard Cole (chap. 5) as well as Gholz (chap. 2) were in agreement that fears here are exaggerated.

10. A point emphasized by Gholz (chap. 2).

11. Chinese foreign policy, as Cole observed (chap. 5), is keyed to improving energy security. Neither he nor other participants were seriously concerned about resultant militancy, however. Kevin Book (chap. 2) and others drew attention to the pressures on developing nations to use whatever technologies were available to them, including coal.

12. A major theme in the remarks of Korin (chap. 2 and throughout).

13. An electrical grid is an interconnected network for delivering electricity from suppliers to consumers. It consists of three main components: 1) generating plants that produce electricity from combustible fuels (coal, natural gas, biomass) or noncombustible fuels (wind, solar, nuclear, and hydro power); 2) transmission lines that carry electricity from power plants to demand centers; and 3) transformers that reduce voltage so distribution lines can carry power for final delivery.

14. While skeptics remain, 98 percent of peer-reviewed scientific papers agree on this point.

15. Climate change is no longer considered merely an issue relating to quality of life and environment, but one that directly affects human and global security, Shaffer, p. 6. The various security implications were discussed in a 2007 joint Triangle Institute for Security Studies (TISS)/U.S. Army War College (USAWC) conference. Proceedings are available online on the Strategic Studies Institute (SSI) website. See Carolyn Pumphrey, ed., *Global Climate Change: National Security Implications*, Carlisle, PA: Strategic Studies Institute, U.S. Army War College, May 2008. Weiss (chap. 4) found widespread support at the conference for his insistence that climate change was a serious problem.

16. See, for example, Gholz (chap. 2); Man-Sung Yim (chap. 3).

17. Cekuta (chap. 5).

18. The importance of infrastructure was stressed at the conference, notably by David Dayton (chap. 4).

19. Douglas Lovelace (chap. 6); James Bartis (chap. 4); and Book (chap. 2).

20. Many wars have been fought over oil: some would even claim that the recent wars in Iraq were primarily driven by our interest in this energy source. See, for example, Alex Roland (chap. 3).

21. Stephen Kelly (chap. 5) stressed the benefits that might be brought by closer relationships with Canada (and Mexico) including the development of pipelines.

22. See especially Book (chap. 2).

23. The advantages of natural gas are highlighted by Vikram Rao (chap. 6) as well as Kelly (chap. 5) and Hegburg (chap. 1).

24. Once vaporized, LNG is highly flammable, which makes it dangerous to transport and, some would say, a possible target for terrorism. Cindy Hurst, "Liquefied Natural Gas: The Next Prize," in Luft and Korin, pp. 271-281.

25. The problems caused by hydraulic fracturing were discussed by Carey King in his analysis of the energy-security-water nexus (chap. 3).

26. See, for example, Book (chap. 2).

27. Geothermal (an energy source not addressed in this conference in any meaningful way but now supplying us with about 1 percent of our electricity) is also a renewable source which might be tapped in the future, especially in the West where there are geysers and volcanoes.

28. If one compares the amount of water that is used by different kinds of technologies in terms of how many gallons of water it would use to drive a vehicle a mile, we find that petroleum takes 0.1; natural gas, about the same; nonirrigated biofuels, 0.3-0.4; and irrigated biofuels like corn ethanol, a staggering 20-30 gallons. See Carey King (chap. 3).

29. These arguments recurred throughout the conference. See for example Michael Roberts (chap. 4), James Bartis (chap. 4), and discussion sessions (chaps. 1, 4, and 6).

30. Bartis (chap. 4).

31. In the United States, nuclear power already provides us with a substantial amount of our electricity: 21 percent in 2010.

32. Bartis (chap. 3).

33. See Steven Miller (chap. 3).

34. On March 10, 2011, just 6 days after this conference, a 9.0 earthquake (Richter scale) and a tsunami hit Japan. The nuclear plant at Fukushima Daichi suffered major damage, reawakening public fears about nuclear energy. The conversation about nuclear energy would likely have been somewhat different had the conference taken place a week later.

35. These problems were discussed by Man-Sung Yim, Bartis, and Miller (chap. 3).

36. Man-Sung Yim's research has led him to believe that countries eager to develop a nuclear industry will avoid moving toward weapons development, as this will undermine their chances of success. Miller is generally more concerned over the possibilities of proliferation. See chap. 4.

37. John Bumgarner (chap. 5).

38. See the presentation by King (chap. 4).

39. A copy of the National Security Strategy (2010) is available from www.whitehouse.gov/sites/default/files/rss_viewer/national_security_strategy.pdf. The goals are laudable but are inherently somewhat contradictory.

40. Rosemary Kelanic (chap. 2).

41. Gholz (chap. 2).

42. See, for example, the remarks by Cekuta and Kelly (chap. 6).

43. Korin (chap. 2).

44. Cekuta (chap. 6).

45. Book (chap. 2).

46. See, for example, Korin (chaps. 2 and 6).

47. Miller (chap. 3).

48. See, for example, Daniel Weiss (chap. 4) and discussion in chap. 6.

49. William Boettcher (chap. 6).

50. The wasteful culture of the United States (among others) has been widely noted. Jared Diamond, for example, claims that the average rates at which people consume resources like oil and metals, and produce wastes like plastics and greenhouse gases, are about 32 times higher in North America, Western Europe, Japan, and Australia than they are in the developing world. See *www.nytimes.com/2008/01/02/opinion/02diamond.html?pagewanted=all*. See also statistics on food waste provided by the U.S. Department of Agriculture, cited in Tristan Stuart, *Waste: Uncovering the Global Food Scandal*: New York: W. W. Norton & Co., Inc. 2009. What lies at the root of this behavior is debated, but in his influential, though controversial paper, *The Significance of the Frontier in American History*, delivered at the American Historical Association in Chicago, Il, in 1893, Frederick Jackson Turner argues that the defining characteristics of the American people, among them an exploitative wastefulness, springs from their experiences as a frontier people.

51. See, for example, Michael Roberts's explanation of "externalities" (chap. 4) and the extended discussion of economic incentives in chap. 6.

52. Korin (chaps. 2 and 6).

CHAPTER 1

ENERGY AND SECURITY KEYNOTE

Alan Hegburg

INTRODUCTION

We begin by emphasizing that the energy economy is one of the three largest economies in the United States, the other two being finance and healthcare. It is very large and affects, as everyone knows, everything from gasoline prices to climate change. It affects the way in which we live and how the economy is organized, so it is extremely important. It also involves a whole range of other issues such as water, climate, dependence on imported oil, and vulnerability to disruption.

THE PAST

Let us take a quick look back at the Saudi oil embargo of 1973 because it has influenced the debate ever since, even though the oil market has changed, the U.S. relationship with the Saudis has changed, and the Saudis' relationship with the world oil market has certainly changed. Moreover, the U.S. Government and Center for Strategic and International Studies (CSIS) have played a role here. CSIS is not an advocacy group. We are not lobbyists. We are a non-profit organization. We are supported by a group of companies, individuals, and others, and we work on a variety of issues that relate to energy markets in all their manifestations. There is a person who does climate, there is a person who does nuclear. We tend to be organized both by fuel and by markets. So if you scratch us, sometimes one person knows a lot more about the nuclear market than he/she does about the oil and gas market, while others know a lot more about the fuel side of the equation than they do about the markets. So we are not psychotic, but certainly multitasking.

The year 1973 was a watershed year for the United States because the Saudis embargoed oil shipments to our country. The reason they did that was that Israel and Egypt were at war. Israel had lost many of its tanks. It had come to the United States to ask for replacements and got them. The tanks were drawn from the stocks in Europe. Recall that in 1973 the tanks were the main weapons that the U.S. Army was going to use to hold the (Fulda) gap. At that time, I was in the embassy in Bonn, Germany, and I can say that the junior officers thought this was a terrible idea. They thought they were being denuded of their equipment and were worried about what would happen if the Soviets decided to attack. Anyway, the tanks went to Israel, and the Saudis decided to embargo the United States.

The embargo failed, and the Saudis were essentially isolated. Given this experience, they will likely never attempt something like this again. So I am not particularly concerned about an embargo in the oil market. The disruption in 1979, which had to do with the Iranian revolution, was much more severe. As those readers who lived through it probably remember, prices became extremely high, rising automatically. There were also many problems with gas lines. In those days my family lived near a gas station, and my kids were young and went into business selling coffee to people standing in line waiting to pump gas.

The United States resorted to odd-even-day and other kinds of gas rationing to ensure that we didn't overdraw our supply.

At the same time, the U.S. administration took a lot of different steps to deal with the crisis, some of which were really quite important because they influence the market to this day. First, it created the Strategic Petroleum Reserve. This is a supply of crude oil stored in caverns on the Gulf Coast. There are some 700 million barrels' worth to be found in Louisiana and Texas. This can be used to supply the market in the event of a major crisis. Second, it took the lead in the creation of the International Energy Agency (IEA) in 1974. It did so to ensure that there would be a collective response to politically-inspired disruption of oil supplies. There have been some 20 disruptions over the last 30 years, most of them small, virtually none of them politically motivated, some of them having to do with changes of governments in a particular producing country.

The Strategic Petroleum Reserve has been used incrementally every so often to dampen down prices. During the invasion of Iraq there was a partial drawdown because the administration rightly feared that the invasion of an oil-producing country would otherwise have driven prices through the ceiling. The ceiling was a lot lower than it otherwise would have been. In fact, the strategic reserve is not something that is used to manipulate the market; it is used for strategic issues.

The IEA is an organization of countries working within the framework of the Organisation for Economic Co-operation and Development (OECD). The organization has a mechanism for sharing oil in the event that there is a crisis. That will probably never be used. The oil to be shared would not be domestically produced oil but rather oil on the high seas. Oil on the high seas has been diverted individually by countries. The United States has done it several times. Individual IEA countries have said that they were having trouble getting oil, and the President and the department have called in companies and asked them if they could divert oil to the country in question. Japan faced the most serious threat after the collapse of Iran because Japan was—and still is—heavily dependent upon Iranian oil imports. Asia in general is also heavily dependent on the Organization of the Petroleum Exporting Countries (OPEC) oil exports.

Also worth noting is the fact that U.S. demand for oil is flattening off. This is because of the greater energy efficiency of our transportation fleet—we now get more miles per gallon. The change in the fuel composition at the gas station is also significant. Ethanol is now being added to the fuel mix, and this is lowering the demand for crude oil. There have also been some structural changes in the energy market in the United States. We are moving toward a service or nonheavy industry economy. That accounts for a fair amount of these changes in demand. I suspect that we will continue the trend we have seen over the past decade and, as we become still more efficient, will see lower energy use per capita.

Let us now review the last 30 years and see where the United States stands vis-à-vis the countries from whom it imports the most oil and natural gas. Of course, the major oil and major natural gas exporter is Canada. Canada is tied by pipeline to the U.S. energy economy. The arguments we have had with the Canadians in the past were over the border price of imports. Today the price is set by the spot market—so the price is whatever the people engaged in the market decide it is going to be.

The U.S. Government is now out of the business of price regulation but was heavily involved in that business 25 years ago. At the time it was trying to dampen down price increases due to price speculation. The emergence of markets—the spot market and the forward market for natural gas and for oil—was the work of President Jimmy Carter. His most important decision was to deregulate oil prices in this country.

Allow me to relate a story on this subject. President Carter made the decision. The oil price deregulation in the U.S. economy took place over the course of 22 months. At the end of this period, President Carter was defeated by President Ronald Reagan, but there was an interim period between the November elections and the January inauguration. I remember sitting in a meeting when President Carter decided to ask President Reagan if he, Carter, could deregulate the remaining 15 percent or so of oil prices that were still being regulated. President Reagan turned him down. Then, of course, as soon as he was inaugurated President, Reagan's first speech was about deregulating oil prices, which he proceeded to do.

But in fact, all the heavy lifting had been done by Jimmy Carter, and all the credit went to Reagan. President Bill Clinton learned from this—after the end of his second term, just before the new government took office, he pardoned a number of people in jail and was roundly criticized for it. I am not taking a stand on whether it was right or wrong to pardon them. The point is that President George Bush could not undo the pardons. In short, if a President wants to do something in this lame duck period, he should do so. As Henry Kissinger would say, don't ask for permission now, ask for forgiveness later. Carter's failure to act on his own explains why Reagan got all the credit.

THE PRESENT AND THE FUTURE

The foregoing events have put us in a good position to look forward in terms of not only the global energy but also the U.S. energy market. We hear a lot of talk about transfer information and a lot of discussion about renewable fuels and things like that. All of that is well and good. We also hear talk about renewable portfolio standards at the federal and state level. There have been many attempts to deal with changing the fuel mix, but, in fact, if you look at the U.S. energy economy, it is about an 80 percent fossil fuel economy: that is to say, it runs on oil, gas, and coal.

Even with a huge amount of effort to bring renewables into the marketplace, scale them up, and make them competitive, the 80 percent number is probably not going to change very much over the next 10 years or so. People may want it to change; I do not deny that. But getting this train to go on a different track requires a huge amount of change in the way in which the policy system is driven. That may be a point of concern because it results in dependency—we shall discuss this later.

But our major concern is U.S. vulnerability. We are integrated into the global oil market, so the refiners can buy from whomever they want. They can find oil on the market that is cheap because it is of a relatively low quality. They can use this low quality crude because they have refineries that are not just geared toward the sweeter ends of the crude oil barrel. That is useful to them, their profit margins, and the economy in general because it means they are not stuck with having to purchase higher-quality crude.

In contrast, the Chinese economy is structured around its crude oil production in Daqing. The oil here is a relatively light, sweet, waxy crude. Chinese refineries are equipped to process that. They haven't been able to convert a lot of refineries, so they import a lot of sweet crudes into China. This is fair enough because we can take a lot lower-quality crudes, so they can pay the higher price for the higher-quality crudes. Part of the ability to weather an interruption of some sort in the crude oil market has to do with the nature of the refinery configuration. U.S. refineries, primarily on the Gulf Coast, but also on the East and West Coasts, have been restructured and can handle a wider slate of crude oils. It gives refiners a wider range of crudes to choose from, and thus they can find the cheaper goods in the marketplace. This is an advantage in the oil market.

We are also taking pressure off the oil market by reducing crude oil imports. Our demand is lessening because of changes in the nature of our transportation fleet and the mix going into the transportation fuel. In the old days, we were the problem in the oil market. We were overpaying for crude oil, and our partners did not like it. Now the reverse is true. We are playing a smaller role in the crude oil market, and that is very encouraging.

Another thing that's encouraging is the development of renewables. These are very important. If you can get them scaled up and into the fuel mix and can do so without the use of heavy subsidies, you have a substitute for crude oil. I do not think this is likely to happen in the near future. Still, it does mean that our oil market is changing for the better.

Now we turn to the subject of carbon dioxide (CO_2) emissions and climate change. CO_2 legislation, most readers know, has been unable to get through Congress. So the real question is, How do you reduce CO2 emissions if you don't have legislative authority? The answer is that you do it with greater efficiencies. You drive efficiency into the energy economy as much as you possibly can, and you find a way to reduce use of heavy duty coal. Coal is a major problem because of the nature of the plants. Many of them are old and need to be retired. The coal industry is worried about CO2 emissions because they will drive up electricity prices, and no one wants to see that. Those in the coal industry are caught in a quandary. If they don't add clean coal into the system, they will be in trouble. But there are problems associated with carbon capture from burnt coal. This has not been tried to scale and has some of the same kinds of problems as spent fuel from nuclear reactors. You put it away and hope someone doesn't break into it 150 years from now and do something bad with it. You have a problem of storage—underground storage. There are a lot of problems, in fact, that center on CO2 and have to do with the way we currently consume and use energy. This is a very difficult issue.

Now a word about domestic natural gas. This is the boom area in the United States thanks primarily to shale gas finds in the area around Fort Worth, Texas, and in Pennsylvania. We have been so successful in shale gas exploration and production over the past year or so that we are reducing our Liquid Natural Gas (LNG) imports. Right now, the LNG that we're importing essentially comes from Trinidad and Tobago, Qatar, and Algeria.

Gas prices are low in the United States, but there are higher prices being rolled into the system. Most companies in the gas industry are on "take or pay" contracts, meaning they are required to take the product from their importers or pay a penalty. But with prices as low as they are, these companies probably cannot compete in the U.S. market. Nor can they compete in the global market because demand for gas on the international market has fallen so steeply. Prices in Europe have gone down substantially, thanks to the recession.

The gas prices in Europe have traditionally been tied to the Russian supply. In the late 1970s when the German government was negotiating with the Soviet government, the U.S. Government sent a delegation to Germany to make two requests of the Germans. Many people thought that the Americans asked them not to buy natural gas from the Soviets. In fact, they did not. They said 1) "Remember Berlin and don't rely on Soviet natural gas for that city." (That was because the Americans had had to bring in coal to Berlin during the Berlin Airlift to keep the power stations running. We didn't want to have the Soviet gas cut off to the electric utilities in West Berlin because there was no quick substitute in the middle of winter.) And 2) "Don't subsidize Soviet gas."

The Germans did in fact agree to the first request. The first gas pipelines from Russia to West Germany did not, in fact, connect to Berlin but rather crossed the border elsewhere and went south. Things have changed since then, of course. The Germans ignored the second request. They indexed the contract prices of gas to the price of oil. Oil trades for about three times the value of gas on an energy equivalency basis, since the energy content of oil is greater than that of natural gas. The price of natural gas should thus be cheaper than the price of oil. When you index natural gas to oil, however, you raise its price to the price of oil and in effect provide subsidies to the Russian producers.

Today the gas market in Europe is changing. A load of LNG now comes into that market mostly in Spain, but also in France and the Netherlands. There is more efficiency in the gas business in Europe. Russians are not selling as much gas as they would like to in Europe. We are turning LNG tankers—especially those coming from Trinidad—away from the U.S. coast. These are going to Europe and offloading their cargos at distressed prices. Thus, if you are smart and not tied into a long-term contract, you can get a real deal.

Our natural gas exploration and production in the United States are having an impact on the Atlantic basin. There are those who would argue that once the Panama Canal is expanded, the Atlantic basin and market, and the Pacific basin and market will actually come together. Right now the price of natural gas is much higher in the Pacific basin than it is in the Atlantic basin. The Pacific market is structured quite a bit differently and does not currently enjoy the benefits of competition. The joining of the markets may prove to be a real boon to the Japanese consumer who is paying high prices for natural gas. That effectively would bring gas-to-gas competition and remove the gas-to-oil indexation that exists there now.

These matters may come across as arcane, but they play a key role in shaping how markets change. These hydrocarbon markets are important, and they're going to be as important 10 or 15 years from now as they are now. Renewables will eventually come into play, but only when they are competitive. Oil and gas are now subsidized, though not quite in the same way other things are subsidized. If they are going to be competitive with alternatives when the price of alternatives comes down—as we hope they will—we are going to have to get rid of those subsidies.

A final note on natural gas. Natural gas in the United States, particularly in Pennsylvania, is under severe regulatory surveillance right now, particularly by the local Environmental Protection Agency (EPA) organizations. There are several reasons for this concern. First, developing this gas requires water and thus above-ground wells. Second, to release the gas, one has to frack (fracture) the granite with water, and the industry drills into granite below the water table to tap small pockets of gas. When this water comes back up, it has a variety of different things in it, including heavy metal and some radioactivity, and in some cases, it can contaminate local well water. So the industry has a disposal issue. It cannot just reinject the water back underground, endangering nearby water tables. Nor can it send the water elsewhere to be cleaned up. Moreover, both oil and gas production calls for heavy use of water, which may be scarce. Once Pennsylvania takes regulator action on these tough issues, we may see similar regulations elsewhere.

These matters are very important because producible gas from these reserves in Pennsylvania and in Texas is about 200 times greater than current gas consumption. Exploitation of this gas could essentially eliminate the need for all of our ship-borne LNG trade and maybe even some of the Canadian trade.

The natural gas system is so much in surplus right now that some companies are actually exporting gas from the United States. This is not gas produced domestically but imported gas which cannot find a market. Thus the United States is now an exporter of natural gas. For now, it sells this gas in very small amounts, seeking permission to export it on a cargo-by-cargo basis. But the U.S. Government is looking at developing a general license process which would permit companies to export gas. Companies would then no longer have to get cargo-by-cargo or country-by-country approval. That step would be actually a very substantial game-changer in the hydrocarbon economy here as well as in the entire Atlantic basin.

QUESTIONS AND ANSWERS

Q: The conservatives in Congress would make the case that the Department of Energy (DoE) has failed in its mission of helping the United States become less dependent on foreign petroleum. Would you comment on the role the DoE plays today?

Alan Hegburg: This issue has been around for some time. I don't know precisely what the numbers are right now, but about 80 percent or probably more of the DoE is dedicated to nuclear weapons, clean up, all those kinds of issues. When the DoE was established, it was a combination of the Energy Research and Development Administration and the Federal Energy Office. But the largest—by far the largest chunk of work and budget and everything else—is done on the nuclear weapons side. It was a military decision. It goes back to the post-war legislation on nuclear weapons in which the military did not have control over certain aspects of the fuel cycle, for whatever reason. There were a lot of military people in the DoE, and there are DoE people that go out to deal with the military.

A lot of the lab system, which is an integral part of the DoE, deals with weapons. The labs tend to be out West where there's lots of room. Those of you who have ever dealt with DoE labs will know that they have diverse interests. There's a very large coal lab in Pennsylvania. There are people who do solar, there are people who do nuclear, there are people who do earthquakes. These are all DoE lab-related activities.

When it comes down to understanding the non-nuclear parts, it is best to go through the assistant secretaries of the various departments and look at their portfolios. There is an international group, which I came out of. We dealt almost exclusively with oil and gas market issues and renewables. It's a small group. Most agencies talk in terms of 500 to 1,000 people. We were less than 100.

There is also a fossil energy group that manages the fossil energy programs. Congress appropriates a fair amount of money to support programs on fossil fuels, renewables, and energy conservation. These are among the non-nuclear programs in the DoE. They're fairly substantial. They have very good staff. The labs report to the assistant secretaries and the under secretary for science who actually does basic science research.Thus if you do get rid of the DoE, that work will have to be done somewhere else.

On the international side, DoE is very active. Let me just give you a couple of examples. There is a series of what are called one-two-three agreements. These are bilateral agreements that the United States requires of any country with which it's going to share nuclear technology. About 19 or 20 countries have these agreements and commit to certain things—probably provision of information—not anything secret but related to proliferation concerns. There are at least four countries in the Arabian Gulf that are interested in nuclear technology: the United Arab Emirates (UAE), Qatar, Saudi Arabia, and Kuwait. There is already a one-two-three agreement with the UAE. I think one is being negotiated with Qatar, but I'm not sure the Quatari are ready to sign off on it. It may seem odd that the

Saudis want to pursue nuclear power, given the extent of their oil and gas reserves. I suspect they are trying to reduce their domestic oil activities because they're extremely inefficient and absorb a lot of capacity that could otherwise sell in the international marketplace. Moreover, their oil is nonrenewable.

The Saudi government has set up a series of labs based on, essentially, the DoE lab system. These labs do different things. CSIS has an agreement with one of the labs. They want to look at what we have, and they want to talk with us. They hold public meetings to discuss a variety of different topics—energy efficiency, renewables, carbon capture and sequestration, obviously, given the size of their oil and gas reserves. The bilateral relationships between the United States and the Saudis and other Gulf states are really quite extensive.

One example of the kind of negotiations that take place is that the Kuwaitis continue to build electric power facilities in Kuwait because electricity is free to its people. When you leave Kuwait for the summer, you leave the air conditioning on and close the door. It's a huge waste. There is actually no price or conservation sensitivity in that marketplace. We've had conversations with the Kuwaitis about that, but of course this is a decision made at the top levels of the government, and so it can't get undone. But the Kuwaitis are coming up against a problem, which is that they only have one site left on the Gulf that can give them water.

The question becomes, do they put a power station there, do they put a refinery there, or do they decide to take steps to make their domestic energy economy more efficient? We've had bilateral conversations with the people that run the system, and they would like very much to find the best answer. We also suggested that they think about power-sharing or other corporate power arrangements with the other Gulf States. On the east coast, Qatar, Kuwait, the UAE, and Saudi Arabia all have large populations. Why wouldn't you collectively build one power station as opposed to four individual power stations? Politics is, of course, a huge impediment, one over which we have little control.

We have greater leverage when it comes to markets. There is, in fact, a whole group of people in those countries who are very interested in changing their domestic energy markets, but again politics intervenes. But in the long run, markets will hopefully trump politics.

Q: You stressed earlier how difficult it is for oil to be used as a politically-motivated instrument of coercion. What, in your opinion, is the situation with natural gas? Can natural gas be used by countries like Russia, for example, as an instrument of coercion against those of our European allies who don't have access to LNG supplies—countries like the Ukraine, the Baltic States, and perhaps several of our NATO members?

Alan Hegburg: This is such a rich topic. Let's put Ukraine aside for a moment since the natural gas pipeline network is not truly integrated in Europe—it needs to be. That would create great flexibility. For example, Spain is not really in that system despite LNG terminals all over the country. Italy imports LNG but in fact also gets Russian gas. One wonders why, being right across from major gas reserve holders Libya and Algeria, Italy is buying it from the Russians.

As I said, we now divert to Europe gas that was coming here, which is actually having an impact on the marketplace. As I said, the Russians in the early days when these contracts were done, were being subsidized such that they were making more money out of their gas by shipping it to Europe than they could anywhere else, and the Europeans were actually doing it consciously. We have told them this is economically crazy. But sometimes politics does trump markets.

Q: Can you talk a little bit about how you see the future of the oil markets? How, in particular, do you think they'll affect our attempt at economic recovery in the future? Gasoline prices right now are increasing and the economy, we're hoping, is starting to recover again. How do you see oil affecting our attempt at recovery?

Alan Hegburg: I haven't been asked about gasoline prices in about 20 years. But you're right, we've seen a dramatic increase in prices, and I'm not sure why, to be honest. We at CSIS don't pay attention to the retail markets very much. Obviously oil, as I said earlier, is a large part of the gross domestic product (GDP) in this country. So if people are diverting disposable income toward paying for oil beyond what they normally would pay, it takes disposable income out of their pockets and clearly has an impact. It is obviously a problem for the vast number of unemployed people in this country who can't afford a whole range of things and have to make tough decisions. There are also a lot of people who are employed whose disposable income is going down. Having gas prices going up by 20 or 30 percent is painful.

How long will it last? Well, the cynic in me says that if I knew the answer I'd take a position in the market and I wouldn't have to answer the question because I'd be so wealthy. I really don't know. We're going into the summer drive season, when refineries go into conversion and start making more gasoline than what they had been making for the winter. That actually could be beneficial if it results in a lot of gasoline coming into the market just when demand is falling off. But I think we have to look for a decline in demand for gasoline, and I don't know if the market has seen that yet or not. That seems to be the only thing in the short term that can actually help bring down gasoline prices.

Another thing that affects oil prices is the locations of gas stations. I worked for Amoco, which was at one time the largest gasoline marketer in the United States. Its principal business was refining and marketing gasoline in major urban areas. It always picked spots that were isolated from every other gas station so you didn't have to compete against the independents. The independent gasoline marketers were always the cheapest. They set the price, and everybody else followed. So Amoco went places where there weren't any other gas stations nearby—that way it could actually charge a few extra cents a gallon. It was a strategy—a location strategy—obviously. So if you want cheaper oil prices, you want to be someplace that is flush with gasoline stations, particularly those owned by marketers who can charge cheaper than the branded outlets can.

I think the gasoline business is going to have to change as we add more biofuels to the mix. Biofuels are coming in, and refineries are not going to make as much gasoline because demand will go down. There are all sorts of factors like this which have to work through the system. They obviously have an influence in a year-to-year market. The other thing to consider is summer restrictions on fuels in urban areas because of smog. There are certain particulates that boil off from the gasoline in the market which can lead to smog. So you have changes in the gasoline mix that also have an impact on the price.

But if you ask me if this crude oil price increase is warranted, I don't see this connection, unlike the old days when you would see crude oil prices go up and it would be translated immediately into gasoline prices going up.

Q: I have two questions. The first is a follow-up to the earlier question about the use of fossil fuels as a weapon. Right now, shale gas technology is a strictly U.S. venture. But there are shale gas deposits in Europe too. Many European companies are taking positions in the Marcellas with a view to gaining the technology. Can the United States use its monopoly as a weapon? The second question relates to a point you made. You said that demand for oil in the United States was flattening—it used to be $21 million a day and it's down to $18.3 now. Are you suggesting that likely it will stay flat and if so, why?

Alan Hegburg: Oil demand right now is about 20 million barrels a day, and the forecast is that it will be about 22 barrels in 2035—that's flat. It's expected rise at a rate of only 0.03 percent, 0.02 percent a year. Why? There are a lot of reasons for that, as I mentioned, like our greater energy efficiency, the fact that we get more miles per gallon, and the fact that biofuels and other alternative fuels are coming into the market. So, yes, our oil demand is flattening out.

Let me go back to the other question about whether or not the United States can use its knowledge of shale-gas technology as a weapon. I think it's a mistake to think just because that technology is being used in the United States, the U.S. Government has some control over that technology. Think of the U.S. oil market as a global market because it is. It's a global market because of the participants in that market. There are a number of companies that are incorporated outside the United States, and they're all operating in the United States. Let's take, for example, British Petroleum (BP), Shell, China National Offshore Oil Corporation (CNOOC), and Petróleos de Venezuela, S.A. (PDVSA). These companies are incorporated outside the United States. They're not located in Delaware or in New York. They operate like U.S. companies, and we think of them as U.S. companies, which is fine. It doesn't mean anything. They're making a contribution. Their stockholders are all over the United States. But the incorporation issue means they're not U.S. companies. Their subsidiaries are U.S. companies. The point is that U.S. companies operate overseas in exactly the same way, e.g., all over Europe.

ExxonMobil was in the shale business in Hungary. It pulled out, why I don't know. But there are obviously shale gas opportunities in Poland and Hungary and probably in the Czech Republic, too. The real question for companies—of all companies of this size and nature no matter where they're incorporated—is how to get a good return for their stockholders. They're looking for investments that actually work. They look at their portfolio and say, Well, if the return here in shale gas is X and our price forecast is Y and it's only six percent here and we can get 12 percent somewhere else, our stockholders will want us to go for 12 percent. So companies like this don't base their decisions on political considerations. They look at investment returns and how to replace their reserves and matters like that.

CNOOC is a case in point. It has bought up a company in the Marcellas. That's pretty typical—what happens is that small companies start the business and when all the hard work has been done and the value proven, the big guys come in behind them. They can pay a premium for the work that's already been done, but then they actually get access to the reserves, which is what they want. The CNOOC was formed as a Chinese offshore company and has dealt with U.S. companies for a long time. A number of years ago, CNOOC tried to buy the Union Oil Company of California (UNOCAL), a California company, and a congressman from the Bay Area opposed it. There was some suspicion about the company at the time, with sinister arguments being made, and they walked away from the attempt. But they are back now, and that's fine. The U.S. Government isn't opposed to them as long as they abide by all the laws that everybody else does.

When you come into the U.S. investment sphere, you're treated like a U.S. company. You do exactly what everybody else does. There's no special treatment for you. So I'm not too concerned about the transfer of technology when it comes to shale. I think the Chinese have shale in the Szechuan Basin. The U.S. Department of Energy has done some studies on the Basin. I think there's probably going to be some drilling if there hasn't already been some. I don't know who's going to do it. The Chinese may do it themselves or they may actually find someone else to do it. Alternatively, someone else may offer to do it for nothing in anticipation of getting the reserves, and the Chinese can actually make the money on the back side of the equation.

Shale gas is going to be distributed throughout the world. I think companies will go for the ones with opportunities that are the easiest and best developed. They don't want to be first to the table. The Marcellas have actually been the ones that have led the way, and the big guys have come in behind them, and that's a normal process.

Q: Early in your talk you sounded quite optimistic about the improvements in efficiency in the United States and the addition of biofuels to the fuel mix. But the United States is a tremendous energy hog in comparison with the rest of the world. Biofuels production has problematic implications for world-wide food prices. How is all of this perceived by other countries? How do they react when we try to give them advice or lecture them about how they should conduct their economies?

Alan Hegburg: I don't know. I've gotten loads of lectures in my life from other countries. People are not shy about telling us what we should do, and what they don't think we should do. I'm reflecting what the forecast is on biofuels and the fuel-consuming fleet, not the question of emerging food shortages. That was a particular issue this year. It may be that this is a constraint and, if it is, there obviously has to be a decision made at the policy level in the U.S. Government.

If you can't put biofuels into the mix, then it seems to me you have to be tougher on the miles-per-gallon standards or flex fuel. Or you have to find some other way to get much more efficient, less carbon-intensive vehicles. That's a policy decision by the U.S. Government. So if it happens, it happens. I'm not defending it one way or the other. I'm reflecting on what the forecasts are for the U.S. crude oil market. I live in the city and rarely drive a car, so from my perspective,

I'd rather see more money in mass transit. There are political questions out there that have to be resolved.

Q: This "question" is more in the form of a comment. I believe that given the price pressure on oil right now and the fragility of our economy that we ought to be seriously considering releasing oil from the Strategic Petroleum Reserve to help put downward pressure on prices. George Bush did that after Hurricane Katrina when the reserve was only about 83 percent full. It's now at capacity. When oil was at $69.00 a barrel, not $102.00 a barrel, it reduced prices within a month, the spot price of oil going down about 10 percent.

In the last year, the increase in gasoline prices has been about 13 percent to the average American family, while income has been stagnant. So it seems to me, given the fragility of the economy, given that the reserve is full, even though we do not have a disruption, we are at a very fragile price point, and we're sending overseas almost a billion dollars a day in petroleum purchases. Now admittedly, a lot of that goes to Canada and Mexico, but still that's money out of our economy that has no multiplier effect. So it's my view that we ought to sell oil from the reserves right now. What is your opinion?

Alan Hegburg: I don't disagree with you. This particular issue has been around for some time, and there have been discussions about it in the past, as you noticed. The reserve was created for a major interruption. If you want to do a draw-down to dampen prices, that's fine. I suspect that it's only a temporary fix, but we can do it. It needs a political decision.

Q: I'm curious to hear some of your thoughts on energy independence and whether that is a national aspiration we should strive toward. It would seem to me that we're interdependent with a lot of different countries for a lot of different things, especially countries like Canada, who are friendly to us. So why is energy any different?

Alan Hegburg: I will refer that question to Jim Bartis.

James Bartis: I believe there are two main reasons that we worry about oil and why oil is different. One is that oil comes from a lot of regions of the world that are inherently unstable. I put together a map of every country that exports more than 200,000 barrels a day of oil. When we looked at that map, there were only three countries where there was a free market in place. Every other country was either a member of the cartel or had serious governance problems or both. So one of the reasons why oil is of concern is the security of the supply. The other reason is that, even though cartels do play an important role in stabilizing prices, they do tend to put the price a bit higher than what we believe would be the price in a totally free market. That doesn't mean we should be foolish about energy dependence. But oil is a bit different than other commodities simply because of those two factors.

Ann Korin: I think you need to define what energy independence means. I'll talk about that later, but energy independence doesn't mean isolation. It doesn't mean walling yourself off from the world market. If you think about it in a correct way, it should mean stripping oil of its strategic status and turning it from a strategic commodity to just another commodity, certainly not walling yourself off from the world market.

Q: I have a question in two parts: (1) how do you perceive the oil depletion or the peak oil concept? and (2) how do alternative or marginal resources like Canadian oil sands come into play here?

Alan Hegburg: A debate over peak oil took place several years ago. Someone wrote a book about it which got a lot of attention. It had an analytical framework, claiming that we had peaked and were headed on the downside of the slope. I'm neither a geologist nor an engineer, but there are a lot of basins that haven't been touched yet that are potentially hydrocarbon-bearing. This suggests that there may be more oil out there. I'm not saying it's ever going get drilled. There are those who say we shouldn't drill the Arctic, for example.

But there are opportunities and there are technologies, and so my instinct is that if we're close to a peak you can sustain that level of production for quite awhile through a variety of different technological applications and maybe economic ones. But that's not a geologic or scientific-based decision. The forecasts that are published by the U.S. Government, the CIA, and others, estimating forward-looking oil production on a global basis, have peak oil continuing to increase, albeit at a slower rate. Putting aside China, obviously, everyone else seems to be in a system with a flat demand going forward or actually a slightly declining demand going forward. Thus, the analytical community, people who have no axe to grind in this game, doesn't seem to accept the peak oil view. But I can't tell you the peak oil is categorically wrong.

Q: You talk about biofuels, and people talk about food, but you can produce biofuel from corn or you can produce it from corncobs through cellulosic work. Before I came to North Carolina, we started a process in Tennessee involving switchgrass. We found that there were about a million and a half acres in the state that were not suitable for growing food but on which you could grow cellulosic grasses like switch grass. If you took two-thirds of that land and produced switchgrass on it and had a reasonable amount of success doing that sort of thing, you could produce about 30 percent of the fuel consumption in the state through ethanol derived from that without impacting the food crop. Another thing we can do now regarding capturing energy from sunlight is to exploit 17 percent of Nevada and probably generate enough fuel for the country. What is your view?

Alan Hegburg: Unfortunately, there's not much oil in the electricity sector, but that's okay. As to ethanol from grass, you're the second person I've heard talk about cellulosic in the past few weeks, so I accept all the things you've said and it may be a way forward. I just don't know. But I've heard a lot of people who I think really know what they're talking about, including you, who have been very, very bullish on cellulosics.

Q: As they would say in the House of Commons in Britain, would the "honorable gentleman" over here tell me how we ultimately achieve a net gain through biofuels when so much energy is consumed in producing biofuels?

Audience: It doesn't actually take all that much energy to produce this kind of biofuel. Take switchgrass. It's a perennial, so you have to turn it over about every 25 years. The most difficult thing they're up against in the Tennessee project right now is the fact that switchgrass is a bit thick, and John Deere hasn't found blades that are tough enough to cut it when it gets a little bit older. But it turns out that you don't have to fertilize very much at all because you can cut it any time of year you want to. The nutrients go back into rootstock when you're doing this and so it's actually not that cost prohibitive. If you're using land that you can't grow food on, it's not a problem in the sense of driving up food prices.

Q: What of the price of harvesting it and extracting the oil?

Audience: Yes, there is a cost to that. But there's a cost to producing oil from any field. It costs to get oil out of the ground, to refine it, and to turn it into a fuel. But I don't think it has to be that prohibitive.

At the end of the day, it's not really very logical to look at the return on energy investment. What you need to look at is the economics. If you look at tar sands—you don't really care how much energy you have to use in order to extract the tar sands. What you care about is whether the cost of the oil that you extract is worth it to you at that price, given the global oil price. The same thing is true of any other fuel. You don't care about the energy input into a lump of coal compared to the energy output in a watt of electricity or anything else. You shouldn't care about the energy input into a cob of corn or stalk of sugar cane. What you should care about is, Does it make sense economically? Take away the subsidies, take away the transport, take away the tariffs, because they distort the science. Remember, we have to deal with thermodynamics. We can never create energy. We can only convert latent energy into usable energy. It's the energy inherent in the matter plus energy of conversion. You can't get around that.

The economics can point you one way, but if you physically have to move goods around the world, other calculations come in to play. You can pay for other fuels to promote trade and the things we use oil for, but the question is how long can you afford to do that? You can relate energy return on investment to the cost of energy in a general sense. They're generally inversely related. Energy-out divided by energy-in is inversely related to cost divided by energy. So you can find the cost trends between any two competing energy sources. The question is, which is more important to us—long-term trend or the short-term trend?

CHAPTER 2

FOSSIL FUELS

This chapter is based on the presentations and discussions of the first panel of the Triangle Institute for Security Studies (TISS)/North Carolina State University (NCSU) Energy and Security Conference. Its purpose was to assess the extent to which our need to secure fossil fuels undermines or enhances our security goals. Anne Korin addressed the dangers posed by oil as a strategic commodity and suggests ways to break the monopoly. Eugene Gholz evaluated the appropriate role of force in defense of our energy security, stressing the need for restraint. Kevin Book focused on the advantages and disadvantages of oil and coal. Rosemary Kelanic launched the discussion period by providing some definitions and commenting on the views of the speakers.

OIL AND GLOBAL SECURITY

Anne Korin

INTRODUCTION

When we talk about oil and security, there are three elephants in the room that we have to deal with. The first is, of course, radical Islam. The second is the Organization of the Petroleum Exporting Countries (OPEC), the oil cartel. The third is actually something good: the rise of the developing world. We have to deal with all three of these issues because they are inherent in looking at oil as a security problem, as something that raises security vulnerabilities.

RADICAL ISLAM

Almost 10 years after September 11, 2001 (9/11), I still don't need to remind anybody that 15 of the 19 hijackers were Saudi, and that wasn't a coincidence. They weren't, let's say, only weeds in an otherwise nice garden. We have a situation in which the Saudi royal family, one family, controls a quarter of the world's oil reserves and has a sort of deal with the devil, which I'm sure everybody's aware of right now. The radical Sunni Islamic establishment in Saudi Arabia is free to propagate radical Sunni Islam around the world. It is lavishly funded by the globe's oil consumers. This includes us, of course.

When we look at radical Shiite Islam, which is primarily funded and propagated by Iran, again we come back to the fact that Iran isn't making its money by selling microchips, or Tootsie Rolls, or anything like that. Thus in whichever direction we look, we always find our path to the world's oil reserves potentially blocked by Islamic radicals. Two thirds of these reserves are located in the Middle East and in North Africa. If one looks beyond the Middle East, a substantial portion of the world's oil reserves in general is in countries in which radical Islam is on the rise. As we see on the news every day, the rumbles and the earthquakes that are occurring across the region bode well neither for the stability of the oil market nor for our economic recovery.

Oil is the second largest household expenditure. It's not just the fuel that we buy at the pump. Every single purchase that we make—food, clothing, everything—is brought to us courtesy of oil. When oil prices goes up, disposable income goes down. When oil prices go up, economies go into recession.

We don't need to talk a great deal about the instability in the Middle East and North Africa because we're hearing about it all the time. There's nothing I could really add to what you already know. We just need to remember that this instability has a number of different facets, such as the Sunni-Shiite conflict, which has been going on for a very long time.

If you narrow your sights a bit further and look specifically at this conflict, then you'll see that Shiites have been oppressed in Sunni countries for many years, despite sometimes being the majority population. Keep in mind that while two thirds of the world's oil reserves sit in the Middle East, it also turns out that Shiites sit on most of the Middle East's oil reserves. Even in Saudi Arabia, the Eastern province of Saudi Arabia where the oil is located, is primarily a Shiite province. The Shiites, of course, are oppressed in Saudi Arabia, but if you think about the conflict in Bahrain and similar places and how that could potentially evolve, it doesn't make you sleep very well at night.

Everybody should be aware of the ramifications of our petro-dollars on the spread of radical Islam. There's no need to get into that. Therefore, let's move to the topic that is much more salient to the context of oil and security.

OIL AS A STRATEGIC COMMODITY

What we need to talk about, really, is where the security vulnerabilities that arise from oil originate. They do not come from the amount of oil that we import. This may sound counterintuitive. However, for the most part, our vulnerabilities with respect to oil would stay exactly the same even if we didn't import any oil. To illustrate this point, think back to 2008. Remember oil prices spiked. At that point, Great Britain was supplying almost all of its own oil. The truckers in Great Britain rioted at that time and blocked the roads. They did so because oil is a fungible commodity; it's got a global price.[1] The price went up for everybody. It didn't matter if you were importing any oil or not.

Our vulnerability thus isn't really about the amount of oil we import. What it's about is the fact that oil is a strategic commodity. To understand what that means, we need to look to history. I titled my last book *Turning Oil Into Salt*, because for centuries salt was a major strategic commodity since it offered the only way to preserve food. Wars were fought over salt. In some cases, countries chose where to establish colonies because of an abundance of salt available. Armies march on their stomach. You couldn't very well march thousands of miles if you didn't have preserved food.

The world moved on salt at that time. Then, with the advent first of canning and then, still more importantly, electricity along with refrigeration, salt became completely irrelevant to world affairs. Does anyone know whether we import salt? How much salt we import? Do you care what the world salt price is? Do you contemplate who the world salt reserve holders are? It's totally irrelevant to you. Unless you're in the salt business, or you're a buyer or seller of salt, you really don't care about any of this. Salt is no longer a strategic commodity. It's just another commodity, despite the fact that we consume more salt now than we ever did before. It's primarily used for clearing snow. But it's no longer important. It no longer shapes the course of history.

Now, oil sits exactly where from a strategic perspective salt did years ago. It shapes the course of history. It shapes relations between countries. Let us be very honest. Would anybody in Washington be using the phrase, "Our friends, the Saudis," in any other circumstances? No.

Thus we need to think about where oil's strategic status stems from. It clearly does *not* stem from our need for electricity. Only 1 percent of U.S. electricity today is generated from oil. Only 1 percent of U.S. oil demand is due to electricity generation. Outside of the oil exporters (in effect the OPEC countries), these numbers are essentially the same globally: 5 percent. Oil and electricity are two separate issues. I'm reiterating this because when you look at polls, 90 percent of the American public believes that solar, wind, or nuclear power will wean us from oil. Ninety percent! This is a huge disconnect.

When you look at politicians running for office, whether they're on the left or on the right, you'll hear calls either to build more solar panels and wind turbines, or to build more nuclear power plants. That'll get us off of oil, they say. That'll end our imports. That'll reduce our dependence. You've heard this. I'm not making this up. It's complete nonsense. Our need for electricity and our need for oil are two totally separate issues.

Oil's strategic status stems from its virtual monopoly over transportation fuel; 98 percent of our transportation energy is petroleum-based. Thus, if we look to the salt analogy, to strip oil of its strategic status, to turn it into just another commodity, we need to break this monopoly. We need to open the fuel market to fuel competition. We shall return to this subject.

Cartels.

Let's hold that thought for a moment and get back to the second part of the picture, which is that this monopoly is married to a cartel. The OPEC oil cartel sits on 78 percent of world oil reserves and, like any cartel, the entire purpose of its being is to maximize revenue for its member regimes. By the very nature of its existence, that's what it does.

In the early 1980s, OPEC produced about 26 million barrels of oil a day. Think of what the global economy has done over the last 3 decades. Oil consumption has grown drastically. Non-OPEC production has grown drastically. You know how much OPEC produces today? Just about the same: 26.8 million barrels.

This cartel has a strategy of deliberately constraining supply to drive prices up. That's what cartels do. In that context, since our foil in this game is a cartel, we need to think about some of the primary policies discussed when it comes to dealing with it.

On the right, it's generally, "Drill, baby, drill." Expand supply. On the left, it's generally. "Americans are consuming too much, we need to drive cars that are smaller." Among the pundit class, you may get calls for gas taxes and so forth. So we have advocates for efficiency and conservation on one side, expansion of supply on the other.

Let's remember our foil is a cartel, and let's see how the cartel deals with these policies. We have a historical event that can serve as a simulation: the oil price spike of 2008. In that year, oil prices spiked to $147 a barrel. Gasoline prices at the pump went up. Consumers are rational economic creatures. In the United States, people responded by driving less. Our gasoline consumption dropped by 10 percent. Oil consumption dropped by a million barrels a day. A similar phenomenon occurred elsewhere—people drove less, and consumption dropped.

What did the cartel do in response? They pumped less. They met several times over the course of 6 months. They cut production by 4 million barrels a day—actually they cheated, so in effect they cut production by 3 million barrels a day. But, in essence, we used less, they pumped less.

Why does this event simulate how cartels might respond to efforts to break their monopoly? You can think of that oil price spike as a proxy for a number of different policies. You can think of it as a proxy for a massive gas tax increase. It's not a gas tax that was paid to Uncle Sam, however. It was paid to Uncle Saud or one of his brethren, but it *was* a gas tax.

So we know how OPEC will respond when the demand is adjusted in that way. We can think of the 10 percent reduction in gasoline demand as a proxy, not an exact proxy, but a proxy for an increase in vehicle fuel efficiency. It is not an exact proxy inasmuch as vehicle fuel efficiency would be longer lasting, but is a proxy in the sense that it is a similar concept. What does OPEC do in response to this sort of static demand reduction? They pump less.

We also know how OPEC will respond when faced by expanding supply. We have to understand that pumping more and using less are essentially two sides of the same coin from the perspective of the cartel. What happens when we expand non-OPEC production of oil? Use tar sands, or anything else? How would the cartel react? It pumps less.

If you look at the last several decades, you'll see that this is a nonlinear relationship because we need to factor in the rise of the developing world, but the relationship is there. So, we use less. They pump less. We drill more. They pump less.

As long as you're playing in the oil field, you cannot manipulate the cartel. As long as your strategies are static strategies and you are relying on static demand reduction or static supply increase, your policies are predictable. The cartel knows how to respond. It might take it some time to erode inventories in the market, but the cartel knows how to respond.

Breaking the Monopoly.

We need to break out of this paradigm by putting other commodities into competition with oil on the vehicle platform. Now, how do we do that? How do we, for instance, monetize the price difference between oil and natural gas today when it comes to transportation? How do we monetize the price difference between oil and coal when it comes to transportation?

Vehicles must be platforms that enable fuel competition. So let's talk about options in the near, mid, and long term. In the near term, the easiest thing to do is go to an open fuel standard that essentially ensures that new cars are platforms for liquid fuel competition. The vehicles should be able to run on gasoline or a variety of alcohol fuels. This capability costs less than $100 a car. You're essentially talking about downloading a different piece of software to a control chip on the car and making the car essentially corrosion resistant. You'll need, in other words, higher grade fittings, because alcohol is more corrosive than gasoline.

What does that do? Let's talk about some of the alcohols that you have out there. Ethanol is one, but methanol is more interesting, specifically because of the natural gas question.[2]

When you ask yourself how you monetize the price difference between oil and natural gas when it comes to transportation, you must ask yourself what people would do if cars were open to liquid fuel competition. They would do just as people did in Brazil in 2008. At that time, 90 percent of new cars in Brazil were flex-fuel vehicles. New cars in Brazil went from zero to 70 percent flex fuel in only 3 years, and they are made by the same automakers who sell cars in the United States. In 2008 when oil prices spiked, consumers in Brazil went to the pump. They compared the per-mile cost of different fuels, saw that unsubsidized alcohol was cheaper than oil and bought it. That year, gasoline became an alternative fuel in Brazil. Brazilians used more alcohol, i.e., ethanol, than they did gasoline.

Ethanol in Brazil is made from sugar cane. We have tariff issues which we can talk about later. They are worth discussing, especially given the need to develop a healthy interdependence with countries in Latin America. Our influence there has been steadily eroding. There are not many tools that we can use to strengthen our relationships. Opening our markets to them is a very good one.

The important thing is that when you have a choice, you compare the economic alternatives and choose what makes most sense from an economic perspective. Let's look at a vehicle that is open to using gasoline and a variety of alcohols. Let's look specifically at methanol, an alcohol that can be made from natural gas and coal, and think about the economics of it. Methanol's spot price today is around $1 per gallon. Methanol has about half the energy of gasoline. So you would pay around $2.50 for the same amount of methanol that you would need to drive as far as you would on a gallon of gasoline. That's including retail markup and everything else. That means that if you could use it in your car, you would buy it unless you don't care about price and don't want to have to refuel twice as often. A lot of people would buy it.

That's what an open, competitive market looks like. The issue is the same if the fuel is made from coal, sugar cane, cellulosic biomass, corn, or anything else. You compare the per-mile economics and decide whether or not it makes sense for you to buy it. If it does, you will buy it.

What does that do to the cartel? We have a bizarre situation today in which a cartel that is full of bad actors and that plays a hugely negative influence on the world stage, controls much of the world's oil reserves. We don't even need to look at the Middle East. We can just talk about Venezuela in Latin America. In OPEC, you have a cartel that controls most of the world's oil reserves, despite accounting for less than a third of global oil production. It is the modulator of the oil market. But if you have an open and competitive fuel market, there will be a lot of energy commodities playing in that market. There won't be a cartel or a group of countries controlling all of those energy commodities. Also, there won't be one single crisis point that brings all of those commodities down, that disrupts all of those commodities. There will be all sorts of different types of crisis points. Each one could bring a given resource down. But you'd have to have a bizarre series of catastrophic failures to bring our transportation sector to a screeching halt.

This is something we should think about since there would be no such thing as a global economy without transportation. That's why oil is so strategic, because it underlies the very fabric of our global economy, of our lives.

Thus liquid fuel choice is step one. Liquid fuel choice is extremely important. As we look toward the future, electrifying transportation is also extremely important. Why? Because on a per-mile basis, electricity is much less expensive than oil. Again, look at the cartel and remember that one strategy it has used periodically is to flush a lot of oil into the market. It does this when it gets concerned that countries are actually going to get somewhat serious about breaking free from its hold. Well, to undercut electricity, they would have to drop the price of oil down to $5 or $10 a barrel. This is not easy to do when you have huge demographic pressures to contend with. Look at the Middle East and North Africa. We're dealing here with sclerotic, corrupt regimes across the board. Many of them, in order to survive, have made a cradle-to-grave bargain with their populations: "We'll supply what you need, just don't storm our palaces." That bargain doesn't always hold, as we have seen, but there's a reason why the Saudis increase entitlement spending whenever instability occurs. It isn't because their princes all of a sudden decided out of the goodness of their heart to distribute more money to the population. It is because they want to keep their heads attached to their necks.

In sum, you can't undercut electricity by manipulating oil price. This means that having electricity in the transportation fuel market will in essence serve to protect the other liquid fuels. That's because whether you look at methanol from natural gas, methanol from coal, ethanol from sugar cane, or even ethanol from corn — and that's with no subsidies, no tariffs — all of them are economic at $50.00-$55.00 a barrel of oil.

You are probably asking yourself how come we're not using these fuels if they are economic. I'll present a very simple analogy to you. Let's say you are lactose intolerant, but you want to start each day with a cup of coffee with some milk in it. If the price of rice milk and soy milk goes to an unaffordable $50.00 a gallon, you still can't put cow milk in your coffee. Essentially, therefore, the cars we are manufacturing today are allergic to anything but one type of fuel. It doesn't matter how economic the other fuels get in comparison to oil, we can't put them in our car today because the engines will not accept the fuel.

Does that makes sense? That's the issue. You can't say that the free market will settle this when we know from economic theory that the market has a very hard time dealing with monopolies and cartels. In effect, by closing its vehicles to fuel competition, the auto industry is colluding with the cartel. It doesn't do this by design, or because it wants to. Nor does it do so because it is engaged in some kind of bizarre conspiracy. It does it out of inertia or habit, call it what you will. But that is what is happening.

During the Q-and-A, we'll talk about that third elephant in the room, the developing world. The major thing to keep in mind about electrification of the transportation system is that it will be a very long time before it happens. It will be decades before we get mass market penetration. We must not look at it as a silver bullet and neglect everything else.

An integrated strategy, including a fully competitive fuel market: that's how we'll break the power of the cartel. That's how we'll reduce the nefarious influence of all of these oil-rich bad actors on the world stage and reduce their ability to destabilize various regions of the world. It's the only way that we can ensure our continued prosperity and economic growth.

It's not a sufficient condition for avoiding national decline but it's certainly a necessary one.

PROTECTING THE PRIZE

Eugene Gholz

INTRODUCTION

The title of my paper as listed in the conference program was "Energy Alarmism and the Strategy of Restraint." An article I wrote a couple of years ago had "energy alarmism" in the title, and I've written a series of articles advocating the need for a grand strategy of restraint for the United States. By that, I mean that we should be a rich country, we should have a powerful military, we shouldn't let anybody mess around with us or threaten us too much, but we don't want to have a busy military. We should be restrained in the way we apply your activist policies around the world. I happen to think those two things marry up together. We don't need to have an activist military, and there are many reasons why we shouldn't have an activist military. Similarly, we should tamp down our energy alarmism.

In the present paper, I shall discuss protecting the prize—oil and American national security—which is the title of a recent article I just published.

DEFINING ENERGY SECURITY

I have a nice picture of a family fueling their car at the pump looking basically happy and feeling good about how easy it is to drive around. In a sense, this is our conception of energy security, how most people talk about it. Casually, we talk about it as access to reliable, cheap energy. That's actually not a particularly useful definition of energy security. I would want to modify it a little bit. I'm going to talk about energy security from the perspective of protection from acute price spikes or from sudden variations in price.

It's these kinds of things that very often link energy to military issues because we think of the military as a crisis response tool. We think of it as something that you can use if there's a political/military disruption in oil markets somewhere that suddenly sends oil prices up or creates a price spike. We think we could use the military to try to tamp down that pressure, either to respond to the spike, or to prevent the spike in the first place.

Consequently, there is this drumbeat that says the reason why the American military needs to be deployed around the world and prepared to be activist and have a forward presence mission is to prevent threats that would otherwise create price spikes—things like a war in the Middle East. People are deathly afraid that there could be a war between two major oil producers which would disrupt the oil market. Then the price would go up, and that would cause a real problem for consuming countries like the United States.

When I talk about energy security, I'm talking about efforts to prevent that kind of sudden political change or sudden price spike. In the preceding chapter by Anne Korin, she was speaking rather of the long-term price of oil, whether it gets higher or lower. She also spoke of the cartel trying to set a higher baseline price, not because of a supply disruption but because, by restricting supplies on a normal day, it can drive the price of oil up. In that case, we would pay a little more for oil than we otherwise would pay.

MARKET FORCES AND OIL PRICE SPIKES

Many things will influence the long-term price of oil: climate change policy; whether peak oil happens; the long-term ability to find oil relative to the amount of growing demand in the developing world. Those are not events that are very easily manipulated by security policy. There is no U.S. military policy that can stop peak oil, if such peaking is truly happening. The military can't create more oil underground. Whether or not we have reached peak oil is a whole other question, which I've written about elsewhere.

Here, the issue is oil price spikes. Let's talk about that. The conventional wisdom goes a bit like this. Suppose there's a supply shock somewhere in the world. Someplace that is producing oil, for whatever reason—there's a war, a revolution, an embargo, or something else—produces less oil. As a result, it suddenly puts less oil into this global bathtub where oil producers are dumping oil into the world market, and oil consumers are sucking it out of different drains all around the world. If somebody's putting in less oil, so this conventional wisdom runs, then it'll be harder to suck oil out of each drain. There'll be less oil in the bathtub, the flow will be slower. Prices will go up and consuming countries like the United States will feel a hit because we have to pay more for our oil.

This story underpins some American and global security policy. The problem with it is that markets adapt, even markets with cartels; in fact, in some ways, *especially* markets with cartels. Therefore, if one country reduces its contribution to the global supply, all the other suppliers around the world have incentives to change their behavior. In fact, they will try to make up the missing supply. Thus what happens is that supply is restored. When somebody goes off production, someone else turns up their faucet a little bit because there's an opportunity to make more money by dumping oil in the bathtub. The result is that no one actually ends up facing a dry pump. When you put gas in your car, somebody else is now pumping oil and the price doesn't go up to a worrisome extent. Supply is restored by natural market forces.

Thus anytime there's a disruption in oil markets around the world, you get a kind of race in the natural market between how much oil goes offline and how quickly substitute suppliers can bring oil back online. This determines how much the price goes up. This is normal market behavior. It doesn't require particular foresight, or especially smart people in the oil business, although there are many such people. Nor does it take someone to plan ahead or develop a complicated policy. Normal market forces create that race.

As Anne Korin and Alan Hegburg pointed out, there's a cartel involved, and what's the whole point of cartels? It's to damp down market forces. The cartel would like to manipulate the price of oil. So your instincts might be to think that, yes, in *normal* market conditions, when there is a supply shock other suppliers will try to increase their output and that would keep prices relatively normal, but that the cartel will undermine this.

But it turns out that cartels are not good at preventing this kind of adaptation. In fact, it's just the reverse. Because cartels limit outputs below what they otherwise would be in a competitive market, they have potential supply lying around. They have a certain level of production capacity that they are not using. They could supply more to the market than they do because the whole point of the cartel is to drive up the equilibrium price.

What ends up happening is that when there's a disruption, all of the suppliers in the cartel—some more than others, but all—have some amount of excess capacity lying around which they can turn on. This can compensate for the disruption caused by the country that has gone out of the market. In fact, the suppliers have more slack capacity lying around than perfectly competitive firms in the market because they are part of the cartel.

It turns out it's actually even worse than that for the cartel because every time somebody goes out of the cartel, every producer in the cartel that has excess capacity wants to be the one who gets to turn on their taps and sell extra oil and make extra money to make up the capacity that was taken off the market. The different producers in the cartel fall all over themselves in cartel negotiations saying, "Let me do it, let me do it." Everybody wants to do it. They have great difficulty making a new bargain, reaching a new political agreement as to who's going be the one that gets to pump more oil. The usual outcome of this is that everybody tries to be the one who pumps the extra oil. This means that more oil comes onto the market than was taken off by the actual disruption because the cartel members can't reach political agreement fast enough to keep up with the dynamics of the market. Thus it's very hard for cartels not to oversupply during supply shocks. From our perspective as consumers, that's great news because it means we're never going to face the actual cutoff, and so the price can go back down.

This is actually what we see when we look at the evidence. If you look at a graph showing the biggest supply shocks that have happened in the past, you can see that there are a couple of times when one or two countries had a big supply disruption.[3] In each case, a declining line shows the disrupted country's oil supply after the disruption. Most of them stay down, but they try to get back up as quickly as possible to their pre-disruption production. But you can see the rest of the cartel, and in fact the rest of countries in the world, the non-OPEC suppliers, lead us back to a 0 percent disruption much quicker, in a matter of a few months, except in the one strange case of the Iranian disruption of 1979. There's a lot of speculation about why Iran in 1979 was different. However, after most disruptions, the market can cover and compensate for them, and we are better off not trying to prevent them. Our actions to try to prevent them are very costly and ill-advised, especially since most disruptions will go away naturally.

I mentioned that there is one case people are very afraid of, and that is when two significant oil producers go to war, a huge disruption in oil supplies is created. In fact, what happened during the war was that there was a sustained reduction in the price of oil because people were overproducing and cheating. Among other things, the countries fighting the Iran-Iraq war wanted the income from oil so they could buy more guns and blow each other up. They needed the money. Consequently, oil production was not repressed. The price of oil dropped, which was good for us. Eventually, Saudi Arabia got tired of everybody cheating. Saudi was really the only country that was trying to maintain the OPEC quota, and you see the sudden price drop in about 1986 when the Saudis said, "Screw you, we're turning on the spigots." The real effect was to disrupt the cartel and create downward pressure. Price shocks disrupt the cartel.

There are caveats to this story. Sometimes OPEC doesn't control the price, and you're back in the normal market response world. There's a lot of debate about when OPEC is more and less important. How well the cartel is functioning varies over time. Sometimes cartel cohesion is high, and cartel members manage to cooperate to restrict supplies. Sometimes the cohesion is low, and they compete more like normal competitive suppliers. But as long as we're at least somewhat in the cartel world, we are less vulnerable to spikes, to supply shocks. Sometimes there's less slack capacity than at other times. That was true in the mid-2000s when the oil price was going way up because Chinese demand was growing very rapidly. During those times, you have a slightly greater likelihood of price spikes.

If you had incredible foresight, if you were God, you could adapt American security policy and say, well, we're going to do a little bit more in the times when there's less slack capacity or when OPEC is not being well behaved. When OPEC is working, then we'll hold off a little bit on security policy and let things evolve. You could modulate.

But we're not smart enough to modulate, and in fact most of our modulations wouldn't work very well. In general, the policy should be to first stand back and let the market evolve because most of the time the market fixes these problems. In those times when after a while the market doesn't fix things, you've got the Strategic Petroleum Reserve. You've got other conscious policies that you can engage. Eventually, if you really wanted to and you were really in bad shape, you could use the military.

POLITICAL-MILITARY EVENTS

I shall comment briefly on three particular kinds of political-military event that might overwhelm a market response to supply shocks. First, what if somebody conquered all of the Middle East? You would then have a better functioning cartel. Suppose the cartel's supply got disrupted or suppose it chose to disrupt its own supply. There wouldn't be any other suppliers in a position to bump up supplies.

Second, what if transportation were interrupted? What if somebody, say Iran, tried to close the Strait of Hormuz, an inviting choke point, through which 17 million barrels of oil a day pass. There are other choke points. What if there was a military interruption of a huge oil producer as opposed to one of the smaller spigots that could be compensated for through another supplier?

Third, what if there was a civil war in a country that is a big supplier? A civil war in Nigeria, which is a moderate supplier, might cut off hundreds of thousands or maybe even a million barrels of oil a day. Other suppliers could easily enough compensate for this. But a civil war in Saudi Arabia could cut off nine million barrels a day. It would be very hard for other suppliers to compensate for that. So you might be concerned with preventing a civil war in Saudi Arabia and might think of using the military to respond.

We therefore must inquire, How real are these threats? Is there a military response?

Conquest.

First, consider military conquest of a major oil producer. The fact that we are probably now at the lowest risk of conquest of one oil supplier by another of any time in history. It used to be that Iraq posed a threat in the Middle East. But one effect of the U.S.-Iraqi war is that we wrecked Iraq's military. There is no doubt about that. Iraq is no longer a serious threat to the other Middle Eastern oil producers.

The U.S. military is very good at coming in from afar using standoff weapons. We don't have to have forward presence to stop one country from conquering another. Our precision capabilities have gone up since 1991, so we can stop a would-be conqueror in its tracks.

Infrastructure Attacks.

The second issue that concerns us is an attack on the oil infrastructure such as disruption of transport. Can somebody close the Strait of Hormuz? In fact, an oil tanker was once attacked and set afire off the coast of Yemen waiting to go into port. Stopped oil tankers are somewhat vulnerable but only somewhat. The good news about this tanker is that, even though it burned for 3 days thanks to a lucky shot, it sailed away. It didn't sink. It continued under its own power. It got repaired. It was not cheap to repair this particular tanker, but that tanker is sailing today delivering oil around the world. It's very hard to sink and disrupt tankers using traditional weapons and military capability. Tankers are very resilient targets. In some ways, they're much more resilient than warships, and there are a zillion of them. Every day something like 22 supertankers pass through the Strait of Hormuz. If Iran were trying to disrupt global oil supplies by attacking tankers, it would have to hit lots of tankers every day on a sustained basis. It would have to continue doing that day in, day out, for months to sustain the disruption. That is a very difficult military mission. It's way beyond the capabilities of Iran, even if the United States stood back and just took the punches, but we all suspect we won't actually do that.

Internal Instability.

The last point is the scariest one. What about internal instability? What if the Saudis had a civil war and couldn't produce any more oil? That would be a real problem. The market couldn't compensate, but what could we do? We would be faced with a really horrible situation where we could swallow the price increase, or we could use the military to try to stop a civil war in a country that is considerably bigger than other countries we've intervened in. There are geographic and demographic reasons why it would be very hard to tamp down the civil war once it started.

So the real question is what should the United States do militarily to prevent a civil war in Saudi Arabia? The best answer is to stay out of Saudi politics. When we have lots of soldiers deployed in the Middle East, particularly around the Gulf, it is a recruiting tool for radicals. Now it goes both ways. There may be reasons we want to do it anyway. There may be reasons we want to have troops deployed there, but we have to understand that the deployments increase the probability of instability in places where we'd really rather there not be instability. It's a very risky and costly policy. Given that the market can handle most of these situations, it's probably a bad idea to pay the costs of actually increasing the probability of the bad outcome, i.e., fomenting internal security that can actually threaten energy security. Thus, in the long view, not being forward-deployed in the Middle

East is probably the best energy security policy that we can adopt.

CONCLUSION

The bottom line is that threats to energy security are probably exaggerated. It's very difficult to generate sustained price spikes through political military supply interruptions around the world. The market compensates. The cartel compensates. This means that the military should play a particularly small role in trying to defend against these contingencies because it's not a practical policy tool. Energy security policy based on the military is a bad response to lots of the energy issues that face us. Better to back away from the military option where feasible, and adopt an over-the-horizon posture in the Middle East, and engage in a policy of strategic restraint.

COAL, CLIMATE CHANGE, AND CONFLICT

Kevin Book

INTRODUCTION

For those involved in the process of geopolitical and economic matters and trying to figure out how they work, for those interested in stimulating and complex relationship between the different moving parts of the global energy economy, and for those grappling with such ultimate matters as death, money, and salvation, it is impossible to avoid the subject of oil. I shall discuss oil in the context of the climate dimensions of security.

OIL

I want to begin by speaking to what an oil analyst like me does. I'm probably less political and more bloodless than anyone who has a dog in the fight here today. That's partly because my clients pay me for only one thing: to look at what the risks are to their investments so they can make money. So far as the larger purpose is concerned, there is little difference between an oil company and portfolio management at a financial management firm, mutual fund, or large hedge fund. They're all concerned with what could go wrong anywhere in the world.

In 2007 I was one of only two of the approximately 35 analysts serving all of the world's institutional capital who suggested that maybe we didn't have *any idea* how much oil there was in the ground in Saudi Arabia. There have been Saudi presentations showing where they explored for oil on the Arabian Peninsula. But in fact the Saudis have barely looked. What they have found is the *cheap* oil.

Anne Korin made the point that the Saudis are capable of pulling back on production precisely because they have very cheap flowing oil. They are also very effective at buffering supply, but what they haven't done is what a for-profit business would do, namely, look for oil that could respond to the marginal demand in the market when the price rises. Thus, in 2007, there were no companies operating in Saudi Arabia that could speak to whether Saudi oil was running out.

My main interest at the time was demand. After all, demand is the one thing that we all understand. We all love energy. I (and the other lonely analyst referred to above) both asked what cost for oil the world could bear. We looked at a couple of different areas separately. The answer we both arrived at was that economies that subsidized energy were starting to spend significant, indeed alarming, shares of their GDP (gross domestic product) when the price of oil per barrel crossed $100. So something was going to have to give.

Here in the United States, where people rightly argued that if there was a higher price for energy, they'd be the ones to pay it, we were also starting to have noticeable problems. We'd been completely indifferent to price for the first 6 years of the last decade. The University of California-Davis Institute for Transportation Studies published a study in 2006 showing that there was no price elasticity of demand for gasoline. It may have been that we were recycling value in our homes, and we had lots of credit. It may have been that we were just really rich at that time and didn't care.

I started looking for the demand response in 2003 and didn't find it until 2008. The way I started looking for it, since it wasn't showing up on the roadways as reduced demand for vehicle fuel, was something we call consumer energy leverage (CEL).[4] Just recently, we prepared a report for our clients, the really masochistic ones who want to read them. It provided an update of what U.S. households spend on electricity, gasoline, and home heating. Again, Korin made a very salient point. It isn't just the price of the stuff that you buy at the pump that hits you when oil prices rise. A large portion of this country depends on propane and heating oil, so an increase in the price of oil affects the average amount of disposable income out there.

To put this into perspective, for about a decade we've been in the 6 percent to 8 percent range of CEL (personal energy expenditures divided by personal disposable income), and in 2007 we got to 10 percent. At that point, we crossed a pretty important line. There's a direct tradeoff once you get there, as Alan Hegburg mentioned, because you can no longer spend your disposable income on higher multiplier parts of the economy.[5] GDP starts to decline because you're putting gasoline in your vehicle at a higher price.

Looking at the intra-year data, we got up to about 12 percent CEL in 2008, before we fell back down to 9.3 percent in 2009. This is on a national average basis. There were locales where it was well above 20 percent in 2008.

This year, with oil currently costing $87 a barrel which was our projection for calendar year 2011, we project that CEL in the United States will hit 10.4 percent, the second highest year on record in a decade. So, if 2008 was a laboratory year for showing how cartels respond to carbon taxes, it was also a laboratory year for showing how the most highly energized, indebted, and energy-consuming economy in the world responds when it has cash in the bank. In 2008 — when we were flush — that response was to drop demand by a million barrels per day in transportation fuels and half a million in the logistics value chain. That created enough slack for even a well-managed cartel to find itself at a price so low that it had to break its own promises.

FOSSIL FUELS AND THE MILITARY

If we are asking what kind of things the military can do to deal with such matters, let me offer two poignant vignettes. In my position, I inadvertently get drawn into public discussions with policymakers, sometimes congressmen who yell at me and say scary things, and sometimes planners in the Pentagon who are doing things called quadrennial reviews.[6]

Of course, I'm just an economist who knows energy. I remember a discussion I had in 2004 about the Joint Battlefield Use Fuels for the Future (JBUFF) program, which was intended to diversify our supply in an ever scarcer world of middle distillates, the stuff that moves jets and material and underpins logistics.[7] The idea was that we were going to look at our shale reserves here in the United States. We were either going to burn it or electrocute it underground so as to pull out the stuff we can put in our jets. We also were going to look at coal-to-liquids technology, a high carbon dioxide emitter.[8] Nazis and South Africans during apartheid found it to be extremely useful for mobilizing hydrocarbons, but very messy from an environmental perspective.

We were looking at all those things as a nation seeking to fuel its military. In the last discussion I had with those same planners, they were talking about how to minimize our dependency on electric power, how to put solar installations on top of buildings, and how to change the cost of operating facilities.

Two things happened during that interval. First, we discovered that there was still a need for oil in the world. In the military forward operating positions, you generally are going to need to have your fuels close at hand. If you find fuels at home of a suitably high energy density, you still have a problem. How are you going to get them overseas to your forward operating positions? You use oil and oil products.

Second, we realized that if the military can lower its operating costs and improve the economics of the folks back home, it can help the folks who are fighting in foreign theaters. That is to say, we have an opportunity here in the United States to help our energy situation. Anything we can do to diversify, which has been discussed pretty abundantly, to manage our utilization, and to improve our conservation, will be useful. It will change the global oil system, at least for a little while, giving it enough time to adapt and enough time for us to get more money so that we can buy the higher-cost resources that can buy us more liberty.

This is one of those beneficial tradeoffs that I keep talking about—the tradeoffs between energy and economy, energy and security.

Coal, Energy, and the Environment.

Oil is capitalism. That's why a financial investor and an oil company think the same way. Coal is core. Coal is crucial. Coal is life.

If you look at who's using which energy in the world right now, the United States is kind of a dinosaur. We're surprisingly coal-levered for a Western industrial economy. That's mainly because our energy economy is a pure market economy. Substantial liberty is extended to the producers and consumers throughout the value chain to find the cheapest way to do things. It turns out that, most of the time, the cheapest energy option is coal. That is mostly because we already have the facilities to consume it.

We have about $2.8 trillion of energy infrastructure in place right now. It's expensive to replace. We use coal because we've been using it for a long time. But we're using less of it as a share of our generating fuels, and we're looking at ways to use it more efficiently when we do. The developing world doesn't have the liberty or the luxury to make this transition, which brings us to the thorny topic of climate change.

Whether or not hydrocarbons cause climate change, climate change is definitely happening. If hydrocarbons indeed make the problem worse, you should probably minimize hydrocarbon combustion. Coal is clearly a hydrocarbon and clearly combusts in a way that yields carbon dioxide. We are not going to be able to control carbon dioxide emissions if the developing world's existential growth toward economic development requires more coal.

There are other interesting aspects of climate security and national security vis-à-vis coal as an energy source.

Here's what's great about energy. It is a real economic marketplace. It is a venue where people maximize value, minimize risk, and behave opportunistically. All the theory you study as an economics student is true, except for one huge "if." Sovereign nations control resources. Indeed, they don't just control resources. They tend also to control trade, resources, infrastructure, and related and supporting industries. Thus the bad news is that whether or not your economic analysis proves valid depends on what the sovereign wants to do. Eugene Gholz made that point very effectively, and it's an important one. A case in point: all of the analysis that my oil price-predicting peers and I did on the global supply system proved to be wrong. Why were we wrong? Because things that happen by sovereign fiat basis can change price very dramatically.

Sovereigns also change allocations and choices. One of the places in the world where they have the most control over energy allocation choices is China. This is one of the places *least* like a market democracy as characterized by a republic run by representative officials. For the better part of the last decade, China did not concern itself with green energy or climate policy. It was in a pure acquisition mode. Sovereignty and supply are one and the same for the Chinese.

In 2003, when I was trying to fathom Chinese energy policy, nobody there would talk to me. But in 2004, all the Chinese diplomats wanted to talk about the growth in China. The difference was that they had succeeded in engendering a reasonably sophisticated supply system. This was growing really fast, and they realized they were going to need to address energy supply to keep their population satisfied, for economic and social reasons. Korin mentioned the most extreme case of this kind of thing. That's when you are an oil-exporting nation, and the amount of oil you sell determines the amount of money you have to placate a population that might not otherwise be quiet and willing to cooperate.

For the developing nations, the choices aren't good. They're going to have to look at coal because coal is stable, and it doesn't take a lot of bells and whistles to turn coal into a functional energy source. Developing nations also have a great advantage, though. It's the advantage of being able to leapfrog over the mistakes that we ourselves have made. Our existing coal-dependent infrastructure is expensive and hard to replace.

The emerging economies have the opportunity to avoid coal-dependency. But there is one problem. It turns out that using green sources, climate-preserving sources, if you want to call them that, or autonomy-imparting sources, if you prefer, requires related and supporting infrastructure. Try to get a wind turbine up to a mountainside without petroleum. Try to make tricholorosilane gas into polysilicon without electric power. There is a bridge between fossil energy and clean energy. We can help developing economies manage that bridge by making sure that they're empowered to support a successful implementation of clean energy. This is actually a huge opportunity. Their countries are the only places we can do it. We are not going to do it here in the United States, not right now. We will eventually. But if we already had a more competitive, more profitable way to funnel solar power into homes, we would be doing it. True, in some places and in some cases, it is more profitable to use clean energy than in others. But if it were economical, it would be widespread here, and we would be using it.

If energy companies like ExxonMobil could find a way to take the massive distribution infrastructure and sales infrastructure they already have and adapt it to sell a different fuel, say, one made from algae or cellulose, or one made from decomposing natural gas into hydrogen, they'd be doing that too. The problem is that they would have to buy new equipment to sell that different fuel, and they don't want to do that in a market-based system when the infrastructure they already have is paid for, and any new infrastructure will require new spending.

Weirdly, nonmarket economies therefore have more opportunities than market economies to adopt alternative fuels. We do business with some of these countries. They actually present us with a wonderful opportunity, provided we're willing to have the same uncomfortable relationships in doing business that we already have when dealing with those same sovereigns. We can help them begin a transition to an autonomy-generating, air-cleaning, possibly climate-cooling energy system.

Herein lies both the hopeful part about coal and the unhopeful part. Coal, where it exists, is useful, cheap . . . and dirty. Cleaning it up turns out to be extremely messy. It's not that I don't like the idea of clean coal, it's just that it really doesn't exist right now. There is a future for the integrated gasification combined cycle (IGCC).[9] It would entail advanced technology coal working at about three times the current wholesale price of power here in the United States. Until that price arrives, the present large coal-fire electric generation companies will remain reluctant to spend money on this technology. Moreover, we as ratepayers in our society aren't particularly excited about it either.

But the answer is to find the way to use coal to encourage greater fuel diversification. Though it is only peripheral to my subject, allow me to touch on the topic of diversifying transportation fuels away from oil.

The natural thing to do would be what the Saudis are doing, which is spending about 30 basis points of top-line oil revenue, in other words three-tenths of 1 percent, on diversifying their own consumption. If they don't use it themselves, they can sell it for export. As Alan Hegburg mentioned, the Saudi fuel oil facilities employed to generate that nation's power are among the least efficient producers of steam in the world. At today's oil price, the opportunity cost to the Saudis, by our calculation, is about the same as the unsubsidized cost of wind generation in the United States. That's effectively a gap closer. The Saudis have an abundant sun resource. They also have a decent amount of silica (silicon dioxide). At some point, solar photovoltaic technologies can close the gap. They are currently about twice that price, with an indifference price[10] of about $130.00 per megawatt hour at today's oil price—the best photovoltaic deployed would come in at about $250.00 per megawatt hour.

There are ways to use fossil energy to pave a pathway to cleaner fuels. We have not even mentioned carbon capture and sequestration (CCS). It works really well. But CCS is one of those complicated issues. If you are going to sequester carbon, you are going to stuff a whole bunch of carbon dioxide underground. So where are you going to put it? The only place it currently makes sense to put it is in an oilfield, and not everyone agrees that that's a good idea. Some people complain that producing enhanced oil with carbon dioxide brings hydrocarbons back into the mix. But it may be a good idea.

But perhaps the best way to capture and store carbon dioxide is something called "beneficial use." This is where innovation enters the picture. We need to find ways to create building materials and byproducts that affordably repurpose the sequestered carbon dioxide in products that can be deployed to greatest benefit. In other words, not deployed here, but in the developing world where they're still creating the structures of urban life. Building materials that capture and store carbon dioxide would be a huge advantage and a huge addressable market for investors.

COMMENTARY

Rosemary Kelanic

Let us try to draw together all of the strands in the panel presentations. Hopefully, I'll also start some discussion between people on the panel and within the audience and perhaps even between members of the audience. Let's begin with a couple of general points that might be food for thought as we're debating these issues going forward.

DEFINING ENERGY SECURITY

There is a huge disagreement about what energy security is. What is the nature of the security threat and who is being threatened? Just on this panel, we've heard some disagreement when it comes to what it is. Is energy security really an issue of price volatility, as Gholz has put it? Is it a long-term price increase issue, as Korin put it? Or is the security issue really something that is more about the nasty by-products of buying oil—like the fact that we end up giving money to regimes that then fund terrorism? That's what Korin talked a lot about.

We've also looked at the question of who is being threatened. Do countries who import more oil face greater threats than other countries? Or do worries over energy security affect everybody about the same? In the field of international politics, when we think about security threats, we are usually not imagining a universal theater. One country threatens another country, and there's a winner and there's a loser. But the oil market is global, and prices are global. If prices go up, everybody is hurt. Isn't everybody threatened about the same? Who are the real victims here? Who wins, who loses?

What exactly do we mean when we talk about the dangers of dependence on foreign oil, as opposed to dependence on oil in general? This was an issue during the 2004 election. The John Kerry campaign wanted to say something that had to do with international security. At the same time, they wanted to avoid mentioning the war in Iraq, which Kerry so inconveniently had been in favor of. Therefore, they made creating a stronger America at home an issue. One of the things they said was, "We want to reduce dependence on foreign oil." This nostrum has acquired a lot of cachet over the past 6 or 7 years, and politicians have hung their hat on it. But it's not totally clear, as Korin said, what difference it makes where the oil is produced.

To what extent are energy security issues really about security as opposed simply to being part of the domestic political discourse or about the security of an officeholder's position? Maybe these issues don't affect a country's security in the sense of affecting its existence—which is the way we usually think about security. Or do they? It occurred to me that over the course of the rest of this conference we are likely to think a lot more about alternative fuels and climate change. Everybody can agree that climate change is a real threat, and this could be a threat to global security.

Then we have the more traditional ideas about security threats which would be war, conflicts, and coercion. For instance, some claim that oil can be used as a weapon to try to blackmail one country to do something it doesn't want to do. These are the sorts of more traditional security threats that we talk about in international politics.

It's not clear what the threat is from a traditional security perspective. It's not clear how bad the threat is, but it seems as if these traditional and untraditional threats are contradictory in a way that may actually be good. When we think of traditional security, we are thinking that our economic, and maybe political, survival depends on the continued supply of cheap oil and cheap energy to keep the economy going. But that cheap oil is the same thing that's actually threatening us when it comes to the nontraditional security threat. To reduce climate change, it seems that what we want to do is try to prevent increased use of fossil fuels. You want to decrease fossil fuels.

So, how do those two things relate? When we face traditional threats, would we be happy about this? Should we be glad if oil prices are going up, or glad if they're spiking and going up and down? It may hurt the economy in the short run, but ultimately if this is a fuel that we need to switch away from in the long run, maybe that's a good thing.

KORIN PRESENTATION

I have a couple of comments on each of the presentations. On Korin's presentation, I think one of the key take-aways, the argument that I think of the most, is that there are all these nasty by-products associated with our dependence on oil. One of these is that we could end up with radical Islamic terrorism. Another is that we could end up with a cartel like OPEC that potentially, has control over our politics, or at least our pocketbooks.

I'll address both of those thoughts and just throw a couple of challenges at you in that regard. First of all, I've always been a little skeptical of the argument about oil funding terrorism, because terrorism is cheap and that's why terrorists use terrorism. It doesn't cost them very much because they don't have a lot of resources. So it's not clear to me that petro-dollars are really that important for terrorism.

Estimates suggest that the amount Timothy McVey spent to bomb the federal building in Oklahoma City was between $450 and $500. The last estimate I saw about 9/11 was that it cost al-Qaeda about $300,000 to fund the 9/11 attacks. That's just not a lot of money. It's not clear that all these billions of dollars in oil are really making a big difference in terms of terrorism funding.

The other point is that where there's a will there's a way. Unfortunately, if the political situation is such that people want to perpetrate terrorist attacks, even if they don't get money from oil countries, they're going to get it from somewhere else.

Korin also talked about the security threat from the Organization of the Petroleum Exporting Countries (OPEC) and the cartel phenomenon. But cartels are not all-powerful. As Gholz said in his presentation, there's a lot of incentive to cheat in a cartel. It may be hard to monitor. You want to sell more oil if you think the price is going to be higher, so in a lot of cases countries do defect from the consensus when it comes to a cartel situation. It's just a huge collective action problem.

Then again, if there are cartel issues—and this is something that Korin discussed somewhat—what happens if OPEC cuts production? Well, economies become a lot less oil intensive. They become a lot more energy efficient, and this blunts the weapon to some extent. You have this adaptation which is actually a good thing, and OPEC countries know this. They know that if they push too far, people will turn away from oil and they'll be powerless. Therefore, they can't raise the prices too much. There is a real limit there, but ultimately they're as boxed in as anybody else. I wonder how much power they really do have when it comes to pricing. Anytime they use their power, they're putting into effect the things that then undo their power, because using it motivates people to modify their behavior and the cartel's power lessens.

Finally, Korin brought up a point about collusion with car companies, and how we can put only oil-based fuel in our engines. We can't use other fuel unless the technology changes, but there is more to it than that in the sense that there's actually a large population of producer gas enthusiasts. During World War II, there was an oil shortage, so what did countries do? They created these producer gas vehicles. All you had to do was add an attachment to your car, and then it could run on coal, peat, or wood. There are pictures in Nazi Germany of people with cars with this crazy little attachment in the front, and they're dumping wood chips into it.

The fact of the matter is that you can look up producer gas vehicles on the internet. There are people that just love this. There's a picture of a Toyota Corolla with one of these attachments on the back, and the guy drives around and fills up his engine by putting wood in the back of the car.

What this all says to me is that it's not so much about the technologies available as it is about will. People can change things if they want to. We can't change people's behavior by changing the options. They have to actually want to run their cars on wood or coal, or they have to be willing to buy cars that are more expensive that can run on both things, or run on electricity. So, it's really more a demand issue than it is a supply issue.

GHOLZ PRESENTATION

Now on to Gholz's presentation. He's basically saying that threats to oil can cause price spikes. Some people out there think that this means we should be alarmed, and that there is a military solution to the problem. But, he says, they are wrong. His insight is that the market can adapt, it can adjust, and price spikes aren't necessarily something that we have to worry about as much as people think we do.

That is true for the most part, except that there is this whole issue of deterrence. If we do have a global oil market that can adjust so nimbly and get rid of these problems with price spikes, isn't that partly because American forces are forward deployed, and we have a footprint in the Middle East? Isn't American military power what's underpinning the peace that allows the global oil market to function? That's deterrence at work. So American forces, deployed the way they are, make it a lot easier to prevent something from ever happening in the first place.

The best analogy I can think of is this. Suppose we live in a neighborhood that doesn't have a lot of crime and somebody says, "We don't really need police forces here because this is a very safe neighborhood. So let's get rid of the police. We don't have to worry because we don't and won't have crime." However, part of the reason why you don't have crime is that the police presence deters it.

Part of the reason we have this peaceful market that can adjust to things is that we know that the United States is involved and is deployed forward to maintain these conditions. If American forces did redeploy, could this have effects that we can't really anticipate because they're deterring things from happening in the first place?

Finally, there's what Charlie Glaser refers to as the "security dilemma" aspects.[11] "Security dilemma" is a rather jargony phrase, but it basically means that countries don't know what other countries want, what they can do, and what their intentions are. They have to worry about what other countries' intentions are in order to protect themselves from those countries. There are all these good reasons why we shouldn't worry so much about price spikes in the market, but countries do worry about it, and given that they do worry about it, they take actions that can cause negative reactions. We might wish that people were a little bit more rational about it and weren't so worried about it, but the fact is that they are, and because they are, this causes other security problems.

BOOK PRESENTATION

Finally, when it comes to coal, it appears that Book is saying that there is an opportunity in developing countries to make useful changes because they don't have an infrastructure in place that's favoring coal.

It strikes me that the political infrastructure is almost more important than the physical infrastructure. One of the reasons why the United States is still using coal, and probably will continue to use a lot of coal, is the existence of entrenched interests. There are lots of people who benefit from the coal industry operating as it is, and that makes it harder to switch away from that fuel. In states like Pennsylvania, West Virginia, and Kentucky, you have major coal interests, and there are pretty powerful people in Congress that represent coal interests and have a good reason to try to block reforms that undermine the coal industry.

You have this whole political apparatus set up that can also undermine change, in addition to the infrastructure issues. In thinking about what can be done in the United States and then in thinking about what can be done in other countries, the reality of entrenched political interests is something that we should also keep in mind when it comes to change.

QUESTIONS AND ANSWERS

Rick Kearney: I want to ask first if any of the panelists would like to make a quick response before we turn it over to the audience. When we turn it over to the audience, I would like you to address your questions to an individual on the panel or to the panel as a whole.

Panelist Responses.

Anne Korin: First, with all due credit to John Kerry, every President since Nixon has been talking about our dependence on oil. Now the fact that politicians talk about an issue doesn't always mean that they have diagnosed the problem properly. As I said, the issue is not one of imports. It's one of oil's status as a strategic commodity. But note that we did accomplish something since the Arab oil embargo. We used to generate a lot of our electricity from oil. Now only 1 percent of our electricity is generated from oil. So we diversified. We're able to switch among different energy commodities when it comes to power generation. We're not able to do that when it comes to transportation fuel.

Second, you'll note I didn't use the word "terrorism." I find terrorism a very bloodless way to refer to the conflict that we're in without actually talking about what the conflict is. It's like talking about the Cold War as a war between missiles or talking about World War II as a war between tanks.

It's not the funding of a specific act of terrorism that is the problem. It's the funding and propagation of radical Islam, and that is fueled by oil. The Saudis have spent some $70 billion on propagating radical Sunni Islam around the world. Ask yourself why the populations of Indonesia and Malaysia, which were primarily very moderate and mild Islamic countries, have become more radicalized. It wasn't because God came down from the heavens and radicalized them overnight. It's because the bulk of the mosques and the religious schools there are funded by Saudi Arabia, and the Imams and the teachers in those schools are trained in Saudi Arabia. If you look at why the Muslim youth are being radicalized in Europe, the people who came from the Middle East and North Africa to Europe came to escape oppressive regimes and to find work opportunities. Why is it that their children and grandchildren have become radicalized? It's because their parents wanted to send them to a religious school to develop cultural affinity. Guess what? If the religious school is subsidized heavily by Saudi Arabia, that's where you're going to send your kid, essentially to be brainwashed. So, I think one shouldn't use the word terrorism without localizing where terrorism comes from. It comes from the religious and cultural climate. It's a climate that develops this notion of war against the rest of the world, be it Christians, Jews, Hindus, gays, women, or anything else. It doesn't come from the air.

Lastly, if you look at vehicles, let's put it this way. Does anyone remember what the price of oil was before 9/11? We're in a different world today. The President of OPEC, the rotating president, is from Iran. But the real power of OPEC has always been Saudi Arabia. The Saudis have spare capacity, and they find a higher price point much more comfortable. When it comes to cars, how many people here are going to retrofit a car? I mean seriously. You want the market to respond in a dynamic fashion. That means people have to be able to make decisions among different fuels on the fly when they fuel their cars. How many people will go and install some sort of odd-looking wood-burning device on their car? Very few.

Rosemary Kelanic: But that's the point. The point is that people won't do it. The point is that we have this problem because people don't want to buy cars that run on electricity.

Anne Korin: The marginal cost of making a vehicle that's capable of running on a variety of liquid fuels is under $100. GM says it's $70 a car. That's less than a rounding error in the cost of a car. Yes. There's a reason that the vehicles that can run on electricity are much more expensive than others, and that is the battery cost. That's why I said it's going to take a very long time for these cars to proliferate. But talking about opening the market to liquid fuel choice is a rounding error on the cost of a car.

Kevin Book: The cost differential between two new car options is $100, but the cost between the car you have and any new car acquisition is 100 percent.

Anne Korin: Yes.

Kevin Book: It's the incumbency of infrastructure that weds us to our fuel choices. Coal and utilities are both stuck in one place, while oil is much more global. Electrons move, though.

We have an electrical infrastructure, and we have electric vehicles today, so it's inevitable that we will come to them if it becomes economic to do so. The only problem we have is that people don't move as well as electrons or oil, and so if you drive into the city from your rich, suburban, happy life, or if you live in the city in your impoverished, squalid home, which you might, you have different utilization flexibility and price sensitivity for the electricity that's being sold in the same place.

We pay for power where we live. We pay for fuel where we drive. These things are policy problems above ground waiting to happen, and if you wait to see what happens when a congested grid charges Grandma in the city a premium, you may say, "We'll penalize the inbound driver." The inbound driver probably can afford to pay the penalty price. All this stuff is a great problem to have, though. To get to the point where there's enough diversity that Grandma's electrical power is getting bid up means that we have a solution that looks like fuel diversity. But right now we're not rich enough to be there.

Anne Korin: If you look at how long a car bought in the United States stays on the road, it's going to be on the road for 16.8 years. At that point, it gets scrapped or sold overseas. So there's a fleet turnover time. That is why you want as swiftly as possible to ensure that new cars sold into the market offer liquid fuel choice which costs essentially nothing. Let's be realistic — we are talking under $100 for this type of modification to a car. When you buy a car, do you look at what's after the comma in terms of the cost?

Kevin Book: It's not $100.

Anne Korin: If you look at the cost of doing it, it's no more than that. There are over 7 million gasoline ethanol flex-fuel vehicles in the United States. GM knows exactly how much it costs. It says it's $70 a car. Ninety percent of new cars sold in Brazil are flex fuel. If you want to make a vehicle that can also run on methanol, it will cost a little bit more, because it's the most corrosive alcohol. But you're looking at a very minor cost. If those cars enter the market, and 2 or 3 years later you're at a point where 15 to 20 percent of the overall cars in the fleet are flex fuel, you've established a pattern for quick growth.

Kevin Book: The problem is the related and supporting infrastructure; you have 169,000 dispensers that have to become ethanol compatible and compatible with any other flex fuels being marketed.

Anne Korin: Let's talk about how much that costs.

Kevin Book: It's not a question of cost, it's a question of whose cost must be paid. It's the distribution terminals.

Anne Korin: Of course.

Kevin Book: The private owner of that terminal has to have an incentive to scrap something that still has economic value in it, and that incentive either has to come from the government or it has to come from the market. It's not in the market right now because the compelling price differential between methanol and gasoline, even if you look at it on the gallon basis, is not economically advantageous.

Anne Korin: Incorrect. The cost of retrofitting a gas pump to serve any alcohol is between $20,000 and $30,000.

Kevin Book: Between $50,000 and $75,000.

Anne Korin: No. The cost of putting in a new one with underground storage is between $60,000 and $70,000. The cost of retrofitting an existing one, which is essentially doing a very good cleanup job on that pump and underground storage, is $20,000 to $30,000.

Kevin Book: For any alcohol? For even the corrosive methanol?

Anne Korin: Yes, for any alcohol.

Rick Kearney: The commentators have been successful in spawning an argument. Are there some questions?

Q: You spoke to the impact that the law of supply and demand plays with OPEC raising and lowering its production. Would you please talk about the impact of speculation on the price of crude oil? In 2008 there seemed to be a lot of speculation in the market that drove the price. We might even be seeing some of that today.

Anne Korin: In 2008 when I was looking at the market and considering speculation, my big question was, Who's driving the speculation? What you have to remember is that, when oil prices are high, we have this huge transfer of wealth going on from consuming countries to the countries of OPEC. Huge amounts of money fill up sovereign wealth funds that are allowed, of course, to play in the market.[12] While in any other industry it is essentially illegal to bid up the price of your own product, I would not rule out the possibility that those sovereign wealth funds were essentially bidding up the price of oil. It's quite possible.

Kevin Book: She's right. It wasn't just because of a malign intent to raise the price of oil, although that might have been part of it. In part, it was because they had more cash than they could place without buying oil futures. It was a weak correlation, but it sure looks as if in March 2008 something weird happened in the differentials. You could actually see what Boone Pickens calls a Texas hedge. That's when you basically buy oil futures with your own money.[13] In part, though, it was because, like any investor, the producers face a portfolio challenge. Sometimes there weren't a lot of other good options. Those sovereign wealth funds aren't necessarily there to bid up oil. They're there to ensure return for their investors. They're supposed to diversify. If you ask me, oil futures were a lousy diversification choice.

Q: Mr. Gholz made an excellent point when calling for active restraint on the part of the political leaders who give orders to the U.S. military. I've been on both sides of a restrained and a not so restrained U.S. military. What about China? China has a burgeoning military force. There's a lot of discussion about how technologically sound that force really is. But China has tremendous interest in securing oil for its economy. Why is it as restrained as it is, and do you see that changing?

Eugene Gholz: That's a really good question. We need to understand the foreign military behavior of China and China as a rising power in global events. The fundamental response that I would have is this: I'm basically an American nationalist. I think the role of American government is to take care of people living in the United States. I want to know how we should deploy the American military to serve American interests. I would be hesitant to accept employment as a strategic advisor to China—China has lots of strategic advisors. But China is worth watching. What I'm basically saying is that I don't care if the Chinese make good strategic choices or bad choices. But, given what China and everyone else is doing, I'd like to do what's good for the United States, and I think restraint is what is good for the United States.

In China, people would say that the U.S. military has command of the commons. As a result, there's not a lot of contention over lines of communication the way there used to be. The Chinese benefit tremendously from this inasmuch as they don't have to protect their own tankers going to China from the Middle East. The United States takes care of that problem for them. To some degree, they get a free ride, benefiting from a U.S. global presence that tamps down some threats.

However, we could be very restrained in our Middle East policy and not engage in lots of activist efforts to manage threats to oil-producing countries, but still maintain our command of the commons, which would continue to benefit China in the way that they're currently benefitting. So even under a policy of restraint where we didn't fight lots of wars in the Middle East, that doesn't necessarily mean we should stop paying attention to sea lanes. But there is some concern that if the United States did less to protect particular oil suppliers in the Middle East, the Chinese might decide that it's in their interest to deploy 50,000 soldiers in the Middle East somewhere just to keep a lid on things. Maybe that wouldn't be good for American interests. Maybe the Chinese have a more mercantilist attitude, and even if it's not good for them, they might decide to restrict oil for their own interests. Maybe they will make a mistake just because they are ideologically confused. That's certainly possible. I think we're a long way from that. That's partly because it's not a good idea for the United States to station lots of troops in the Middle East, and I rather think Chinese strategic advisors might think the same thing. The market works fairly well. We don't need to station troops there. Why the Chinese would make such a mistake, I don't know. That's one interesting question.

The second thing is if that if China did deploy lots of soldiers to protect individual suppliers in the Middle East, and did try to establish a forward presence, it wouldn't solve their strategic problem. The fact is that we have command of the commons and so no matter how many soldiers they deploy in the Middle East, they still couldn't ship the oil to China. They have to get it to their Chinese ports. The fundamental Chinese security problem is thus not how to develop a clever policy which allows them to use their military to prevent a civil war in Saudi Arabia, but how to develop a military policy that guarantees the transit of oil to Chinese ports. This gives the Chinese a naval bias, not a bias toward ground forces in the Middle East.

There's some evidence that that's how they think and what they're doing. They are getting rich, and they are getting powerful militarily in a way. But they're a long way from being competitive and in command of the commons. That is a strategic problem for a couple of decades from now. Not for today. America's strategic problem today is how to remain restrained. Then we can handle the Chinese.

Q: I have hands-on experience with some military command and control issues. But more to the point, I have experience with alternative fuel vehicles. For instance, you speak of methanol vehicles. I had one in Southern California. Unfortunately, the number of places I could buy methanol for the vehicle was zero. Also, I've been working with natural gas. I just finished driving across country in a natural gas vehicle which is out in the parking lot, and I assure you it's like driving across country in a gas vehicle in 1906.

But there are ways to make things better. For instance, what we can do with the existing fleet which, as has been pointed out, will take years to get rid of? The only alternative fuel that existing gasoline internal combustion vehicles can use is natural gas. A lot of countries in the world, Argentina, for example, are using it. I've checked into it and have price quotes from Argentine manufacturers for conversion kits to natural gas for $800.00 a copy. In addition, you have to get the tanks for it. But you can do it. It is possible. They did it. It has been done in other parts of the world. Argentina has the most. Pakistan and Iran are next. The United States ranks right in there between Egypt and Armenia. But it can be done.

But we do have to make changes. Part of the impediment is government regulations. For instance, it is expensive and arduous to meet the Environmental Protection Agency's requirements for kit-car certification. But it can be done. Converted cars are usable. I drive in North Carolina in natural gas vehicles, and I do it out in California. They're quite usable. They look and drive just like any other vehicle.

But if you want to address energy security problems, you have to address quickly what are viable alternative vehicle energy sources in the United States. The answers are few. Electrics can be produced and are ideal for some of the driving environments which most people have to contend with.

Anne Korin: I would like to reply. The speaker actually illustrated very well why we don't want the government to say you need to go from one fuel to another fuel. A dedicated alternative fuel vehicle poses a huge chicken-and-egg issue.

What you want is a vehicle that's essentially an open platform that solves your chicken-and-egg problem. If you can use a variety of fuels in the car, you can continue to fuel the vehicle with gasoline while allowing time for the infrastructure of other fuels to catch up. You as a consumer are not inconvenienced. You can fuel your car with gasoline if you need to. When there's an economic case for you to fuel with something else, you will do that automatically without prompting from the government. But you're not going to see the fueling infrastructure we were talking about catch up until some 15 to 20 percent of the cars of the overall fleet—not just of new cars, but of the overall fleet—are flex fuel. If you are a fuel station owner who has 10 pumps, it is at that point and only at that point that there is a business case for you to either retrofit one of your pumps to serve an alcohol fuel or to put in a new pump. The cost is not prohibitive.

If the station owner goes to alcohol, there is a $50,000.00 tax credit. But nobody's going to use this tax credit until there is actually a business case to be made. We can learn a lot from the example of Brazil. Brazil started with a poor policy decision to dictate that Brazilians should shift from 100 percent gasoline to 100 percent ethanol cars. What happened? Oil prices fell. Sugar prices went up, and people said, "We don't like this." Then they realized that with cars that offer choice, competition enables drivers to select among the different fuels.

Q: I understand that it doesn't take many new stations to have a rather good infrastructure. For instance, to put an alternative fuel station every 100 kilometers of interstate highway in the United States would take only 750 stations. This may seem a lot, but it is a far cry from the hundreds of thousands of gasoline stations which are out there now.

Kevin Book: The problem with that is that you need to drive 100 kilometers to get to the station, which isn't a problem if you've got a really big tank. The vehicle you want to retrofit first is the long-range, fixed-route, heavy-hauling vehicle (18 wheelers). If you're going to pick a standard for internal combustion engines, I'm all for sticking with the standard we've got. Fuel diversity in effect would probably not result in burning natural gas with a retrofit kit, but rather in using synthetic fuels made from the natural gas, like methanol.

Anne Korin: Why not methanol?
Kevin Book: It's an infrastructure question. . . .

Q: I have a question for Mr. Gholz. When we look at OPEC, we don't see a very cohesive cartel. In fact, we see maybe three or four countries, all of which are in the Gulf, that are setting prices. Iran is not one of them. When you look at OPEC's future and the fault lines in that future, and in particular some of the projections with regard to Iraqi production, what do you see?

Eugene Gholz: It's very hard to see inside the dynamics of the cartel. We don't really have a good feel for where the oil is or how much there is. Production levels and pricing decisions of OPEC members are state secrets. The information is protected just like our own classified information. So it's hard for us to know what's going on.

From the outside, we model what the behavior of these countries looks like. We see observed prices in the market, and we try to infer what their decision-making criteria must be. This isn't the most reliable process in the world. I think it's one of the reasons I would prefer not to make American policy depend on getting exactly right the model of pricing behavior in the market. I'd rather just say look, the cartel sometimes works better, sometimes works worse. Let's just have our policy take into account that there is a cartel.

As to the future, it does look like there's a core of OPEC that really leads the cartel. Some countries have much more slack than others because they're the core. Others look much more like Norway, which is not a cartel member and basically produces in max mode. That's to say that they produce as much as they can at the current price so as to make as much money as they can. They look like normal market players.

Countries enter and exit the core. You say that Iran doesn't look like it's in the core right now. Iran has traditionally been one of the most hawkish companies on price. They like high prices. They're just bad at implementing their desire to be part of the high price cartel within the core of OPEC. Again, different countries will behave differently. If Iraq starts producing a lot, it's got to look like Saudi Arabia in some way. They've got to come to a deal with the Saudis to make the cartel work, and I see no reason to believe Iraq/Saudi negotiations are going to be easy.

I don't imagine that the cartel is going to routinely function well, but I do imagine that it's going to have some routine functioning, and that's what I would base projections and policy choices on. Let's say there is some insulation against shock. In fact, if Iraq gets big, there will be more insulation against shock because Iraq will be better at insulating against the Saudi civil war kind of shock and because Iraq has the potential for a lot of slack capacity and the potential to use export routes that are not through the Gulf. The closing of the Strait of Hormuz is the pivotal scenario, but as long as the Strait is open—which I think is essentially inevitable—then you're just involved in the international politics of negotiations and shocks and responses. It's Iraq, it's Saudi Arabia, it's Iran. Everybody's in the game.

Q: I just don't know how Saudi Arabia produces.

Eugene Gholz: Nobody knows, right? You mean economically?

Q: They were in trouble, and there's a good reason. There's a good argument to be made that it was the very low prices that started the Saudi's domestic terrorism problem.

Eugene Gholz: For sure.

Q: Not the high prices.

Eugene Gholz: Or the sustained American decline in Saudi. There's this other factor that after 1991, we left a lot of troops there which radicalized a lot of people that might not otherwise have been radicalized. So it's not high prices.

Q: The oil was $10 a barrel.
Eugene Gholz: Absolutely.

ENDNOTES - CHAPTER 2

1. Fungibility is the property of a good or a commodity whose individual units are capable of mutual substitution. Examples of highly fungible commodities are crude oil, wheat, orange juice, precious metals, and currencies.

2. Methanol is an alternate fuel for internal combustion and other engines, either in combination with gasoline or by itself. In general, ethanol is less toxic and has higher energy density, although methanol is less expensive to produce sustainably and is a less expensive way to reduce the carbon footprint. It may be made from fossil or renewable resources, in particular, natural gas, coal, and biomass, respectively.

3. The graph shown at the conference is not included.

4. Consumer Energy Leverage (CEL) can be defined as the fraction of disposable personal income (DPI) that consumers spend on their electricity, home heating, and gasoline/diesel.

5. A change in price level has the potential to affect many different parts of the economy either in positive or in adverse ways depending on how the effects happen. The multiplier effect takes the original wealth, international, and interest rate effects and amplifies them.

6. The *Quadrennial Defense Review* (QDR) is a legislatively-mandated review of Department of Defense (DoD) strategy and priorities.

7. A general classification of fuels that includes heating oil, diesel fuel, and kerosene.

8. To transform coal into a liquid fuel, coal is mixed with oxygen and steam at high temperatures and pressure to produce a gas. This gas is then reacted in the presence of a catalyst to produce a synthetic oil. However, it takes a lot of energy to loosen up the carbon bonds in coal. Second, all that *energy* use results in the emission of a lot of carbon dioxide—the most ubiquitous greenhouse gas causing climate change.

9. An integrated gasification combined cycle (IGCC) is a technology that turns coal into gas — synthesis gas (syngas). It then removes impurities from the coal gas before it is combusted and attempts to turn any pollutants into reusable by-products.

10. The value of something is whatever we are (just) willing to give up for it. Two things have the same value if gaining one and losing the other leaves us neither better nor worse off—meaning that there is no difference between the situation before the exchange and the situation after the exchange.

11. Charles L. Glaser is Professor of Political Science and International Affairs and Director of the Institute for Security and Conflict Studies at the Elliott School of The George Washington University.

12. A sovereign wealth fund (SWF) is a state-owned investment fund composed of financial assets such as stocks, bonds, property, precious metals, or other financial instruments. Sovereign wealth funds invest globally.

13. A "Texas hedge" is a financial hedge that increases exposure to the risk one intended to mitigate.

CHAPTER 3

ALTERNATIVE ENERGY: NUCLEAR AND WATER

The panel on which this chapter is based was organized as a "conversation" between a moderator and panelists. Its purpose was to highlight the ways in which existing and future energy technologies affect security at the human, national, and collective levels. This panel focused on nuclear power and the use of water as a direct and indirect source of energy. The moderator was Alex Roland. Panelists were Steven N. Miller, Man-Sung Yim, Carey King, and James Bartis.

INTRODUCTION

Alex Roland

I'm a historian from Duke University, specializing in security studies and the history of technology, so I know something about energy, but probably not as much as all of the rest of the panelists.

These paired sessions are on alternative energy sources. This panel, the first one, will discuss nuclear and water, considering water as both a direct source of energy and then as a supplementary resource in more complicated energy delivery systems. The second panel will do wind, solar, and biofuels, including cellulosics. In both panels, we will also be looking at the military implications of these sources of energy. We are the non-fossil group.

Each of our panelists will provide a brief opening comment on this topic from his point of view. When they've finished, I will offer a few questions to direct us to some of the major issues that might come up under these topic headings.

NUCLEAR PROLIFERATION

Steven N. Miller

My mission is to introduce the subject of nuclear power and nuclear proliferation. In recent years, the world has grown much more interested in nuclear power. In fact, we're in the early stages of what some people have termed the nuclear renaissance, which has to do with the growing appetite for nuclear power around the world. In the last few years, 65 countries that don't now have nuclear power have approached the International Atomic Energy Agency (IAEA) and formally expressed an interest in pursuing nuclear power. Some of those are pretty far along; some are barely a gleam in the beholder's eye; many of them will never have their nuclear dreams come true. Thus the 65 number is not a true barometer of where we're going to be heading in any foreseeable future, but it is an indication of the growing appetite for nuclear power around the world.

Meanwhile, we are seeing two related but separate phenomena, each of which raises its own issues. One is the substantial expansion of nuclear power in some places where it presently exists, particularly China, Russia, India, and South Korea, all of which have very aggressive nuclear power construction programs. China today has 24 reactors under construction. The

United States has started one new reactor since the mid-1970s. The Chinese broke ground on three new reactors in December 2010. The Koreans are also very aggressive and have had a building program over the recent years that puts us to shame.

The other issue is the spread of nuclear power to places where it doesn't presently exist. The leading edge of this is found in places like Abu Dhabi in the Middle East or Vietnam in Southeast Asia. These are countries that have chosen vendors, signed contracts, negotiated deals, chosen and characterized sites, and spent money. Things here are getting real as fast as they can get real in the nuclear sector, which operates within a very long timeline environment.

For example, Abu Dhabi has made a decision to have four nuclear power reactors. They have chosen their vendor, a Korean electric power corporation. They've signed a contract for 40 billion dollars over 20 years divided between reactor construction and training. They aspire to have the first reactor connected to the grid in 2017.

Up until now, there has been zero nuclear power in the Middle East—zero. The Israelis have a reactor, but it's related to their nuclear weapons program and not used for power. In the future, Abu Dhabi is going to have nuclear power. The Egyptians and the Jordanians are both a little bit behind the United Arab Emirates, but have made their decisions and are moving forward. A similar story can be told about Southeast Asia, where Vietnam is leading the charge and aspires to have a reactor connected to the grid by about 2020 or 2021. Its long-term aspiration is to have 14 reactors built by 2030. It is a very substantial program.

There are two parts of the world, the ones I've mentioned—Southeast Asia and the Middle East—where the appetite for nuclear power is nearly universal. The only states in the Middle East that have not formally expressed an interest in nuclear power are Syria, Iraq, and Lebanon. Iraq has informally expressed an interest; so has Lebanon. Syria, on the other hand, never approached the IAEA, but did, in fact, try to buy a reactor from North Korea — a modest hint of interest in nuclear power and perhaps other things.

Many of these nuclear dreams won't come true, but enough of it will such that in the future we will be living in a different world than in the past. Up until now, nuclear power has been confined to about 30 countries, and limited to 440 reactors, 104 of which are located in the United States. There's been almost no growth in the global fleet of reactors since the mid-1980s. That is now changing. There are now 61 reactors under construction in 15 different countries and one now under construction in the United States for the first time in several decades. However, we are mostly sending them out to other countries, and that has its own implications.

Why is this trend occurring? In answering, I will preface my remarks by broaching two important considerations. One is the long lead time associated with the development of nuclear power. Typically, it takes about 10 years at least from the first gleam in the eye to the first kilowatt hour of electricity. Depending on the regulatory context, it can take even longer than that. Two, we're talking about long-lived assets. In earlier generations, the expected life span of a reactor was about 30 to 40 years. Now, as reactors are coming to the end of their service life, we're discovering that, for a relatively small sum of money, a few hundred million bucks, you can extend their life span for another 20 or 40 years. There is no better cash cow for a utility than a reactor that's already paid for. So now we're thinking in terms of 60- to 80-year life spans. It takes 10 or 15 years to get to where you actually have a functional reactor connected to the grid and a 60- to 80-year life span after that. It changes the whole way in which you think about the arithmetic of making this investment. It's within this broad context that people are making bets and guesses about the future.

Though there are a number of reasons why we can expect growth in the nuclear sector, every one of them has complications and nuances which I'll simply mention. There are worries about prices, fossil fuel prices. There are expectations that perhaps we're entering an era of high fossil fuel prices that will be made permanent because of a changing market structure and greater demand caused by the entry of China and India to the market. Such thoughts are animating policymakers in certain parts of the world. There are worries about threats to access for similar reasons. There is rapid growth in demand for electricity in a lot of places: Iran, Abu Dhabi, and Dubai, for example. As a result, policymakers are casting around for all possible ways of satisfying this demand, including the use of fossil fuels to generate domestic electricity. Why, then, are places like Saudi Arabia, Abu Dhabi, and Qatar pursuing nuclear power? It's because they envision a future in which fossil fuel is so valuable that it's cheaper, or more prudent, to generate electricity with nuclear power and preserve their fossil fuel for use as an export commodity.

In the Middle East, desalination is a nontrivial consideration for some of these states. Strategies of energy diversification, of course, have very strong links to global climate change. There is this large-scale energy producing asset, nuclear power, which does not generate greenhouse gases. Also, of the alternatives to fossil fuel, it is the one whose scale can be accurately controlled. The reactors that the Koreans are selling to Abu Dhabi are 1,400 megawatts. They're buying four of those. Moreover, possible carbon taxes, or other artificial increases in fossil fuel prices, make people think that in the future the economics of energy may be more favorable to nuclear.

In a number of places, despite the fact that nuclear technology is an 80-year-old science and 70-year-old technology, it's associated with modernity, modernization, and keeping up with the Jones's. There is both a symbolic and an economic dimension to this. When you talk to Iranians, they say that acquiring nuclear power is an integral part of their quest to join the modern economy. In some places, there's a status connection to this. Nuclear power becomes a kind of national project. I think this is true in Abu Dhabi. It's true in some other places like Iran. It's not the decisive reason, but at the margin it becomes important because in some places it removes the nuclear consideration from the constraints of commercial calculation.

Another variable may be the waning of the Chernobyl effect. The story of the global fleet of reactors for several decades went like this. There was steady growth in the number of reactors that the world invested in until it flattened out in about 1986, and it stayed more or less flat; only a tiny uptick ever since. Now we've had several decades without a further catastrophic incident and the generations are changing. (Note: This conference took place just 6 days before the nuclear disaster at Fukushima on March 10, 2011. This event has considerably altered thinking about the viability of nuclear energy.) But there are billions of people on the planet who weren't alive in 1986 and have no memory of the Chernobyl catastrophe. That's been a kind of liberating variable.

There are a number of reasons, not all mutually exclusive, that cluster together in various combinations in each national capital where nuclear power is being contemplated. What, then, are the security implications of this potential nuclear renaissance? This is not the first time we've been anticipating one. In the past, it hasn't come true. I believe that there will be change enough so that we will be living in a different world, even if many of the nuclear dreams that are currently out there don't ever come true.

I'll make three points regarding the security implications. One implication is *not* that, and I emphasize the *not*, the geopolitics of energy will be transformed. Nuclear energy today accounts for approximately 16 percent of global energy consumption. Over the next 20 or 30 years, that percentage may well fall rather than rise. That is not because a nongrowth of nuclear power, but because everything else is going to be growing faster.

One also has to take into account nuclear plant retirements, because we haven't invested in nuclear power plants for a long time. Of the current global reactor fleet, 80 percent is 20-years-old or older. Accordingly, a lot of plants are going to be retiring, even with service extensions. Therefore it's still mostly going to be a fossil fuel world or some other. Nuclear power is not going to exempt us from those kinds of considerations. If you're worried about the Persian Gulf because of oil and gas, in 20 years you're still going to be worried about the Persian Gulf because of oil and gas. The nuclear renaissance is not going to change that.

There are two security implications that we do need to worry about somewhat. One has to do with the security of nuclear installations, which links directly to the question of nuclear terrorism. At first glance, this seems like an odd one because the necessity of nuclear security is obvious, and most people take it for granted.

The questions arise when you get down to the next level and ask what it means to have a secure facility? It turns out that answering that question is very tricky. In the United States, we're 6 decades into our nuclear industry. We still don't have a clear answer, and the Department of Energy (DoE) is having a big fight about it right now. What is the design basis threat against which a utility needs to plan? What are the standards to which it ought to be held accountable in terms of provision of security? Provision of security is expensive on a pure cost basis. There's no profit or revenue stream associated with it from a utility point of view. This is a deadweight loss. So you want enough security, but not too much. How do you define that line? How many simultaneous points of intrusion? This is a big issue.

A pivotal consideration turns out to be whether you assume insider help or not. It is much tougher to deal with potential intruders if they have insider help. In one of the most serious incidents we've had, the Pelindaba intrusion in South Africa, they had simultaneous intrusions at two different points in the security perimeter, by intruders who clearly had inside help because they knew exactly where to go and exactly how to get there in the facility. How heavily armed are the intruders? How many intruders? Depending on how you answer those questions, your security is or isn't adequate. Again, around-the-clock guards are extremely expensive, even disregarding their required training and armaments.

The second security implication involves proliferation. This does not have to do with the spread of light water reactors. Light water reactors themselves pose no particular proliferation threat. The threat is entirely related to the associated fuel cycle elements: the enrichment of nuclear fuel at the front end and the reprocessing of spent fuel at the back end. The first produces enriched uranium which, if it's enriched enough, becomes a weapons-usable material. The reprocessing at the back end extracts metallic plutonium, which is also a weapons usable material.

In sum, the proliferation implications of the spread of nuclear power depend very heavily on the fuel cycle choices made by aspiring nuclear power states. If they choose the paths that we would prefer, which is to say forsaking the worrisome fuel cycle elements, the proliferation implications are circumscribed.

However, in Iran we see an example of a state that's doing the opposite. In South Korea, we see a state that's adopting reprocessing of fuel, as a "waste management" strategy. In Egypt, Jordan, Saudi Arabia, we have states that are expressing vague, long-term interest in uranium enrichment. It's not at all clear that the world is predisposed universally to sign up to the technology path that would be most reassuring to us.

If we fail to manage that process carefully, we could end up with a replication of what I call the Iran Problem, which is dual-use technologies that have equal weapons and power-generating capabilities. This will mean that in any place where there are fuel cycle capabilities and suspicious intentions, we will have protracted proliferation crises of the Iran type.

PATHWAYS BETWEEN CIVILIAN AND MILITARY NUCLEAR POWER

Man-Sung Yim

I begin with an important question: Is there a relationship between nuclear power and nuclear proliferation? There are currently about 30 or 31 countries operating commissioned nuclear power plants. Six countries currently own nuclear weapons. Of the 23 countries that at one time or another explored the idea of developing nuclear weapons, only 4 of these have become nuclear weapons states: Israel, India, Pakistan, and North Korea. What does this tell us about the pathway between civilian nuclear power and nuclear weapons development? It's actually rather hard to say whether it has a direct relationship or not. One way to find out is to look at the history of nuclear proliferation behavior. There are precedents for this type of work. Steven N. Meyer did a salient study in 1984. Little was done for a few decades thereafter. Then suddenly in 2004, Chris Way and Sonali Singh wrote a paper on this topic. They were followed by a number of scholars including Matt Fuhrmann, Matt Kroenig, and eventually Scott Sagan, who all wrote papers on patterns of proliferation.

I have been engaged in an effort with my colleagues to create a model that might help us predict proliferation behavior. Such a tool, we think, might prove useful for policymakers. To do this, a model must consider the kinds of factors that make a country want to proliferate and the various dynamics at work within a country's specific situation and capabilities. It must also be applicable to specific scenarios chosen by a country.

There are some potential pitfalls and limitations in the use of models which we are trying to overcome. Among these is the difficulty in gathering the relevant information. For example, we want to examine the motivations, the dynamics, and requirements of nuclear proliferation. We want to look at both the supply and demand factors. To that end, we are developing our own database. Some of the data comes from Sonali Singh's work—he used the Correlates of War Database—and Doong-Joon Jo and Erick Gartzke's work.[1] We also relied on the open source data from the IAEA. We have collected five different sets of data on 1) economic development, 2) security environment, 3) international status, 4) political development, and 5) nuclear technology capability, and the national status vis-à-vis nuclear nonproliferation norms. There is nothing sensitive about these data. We've tried, and we actually continue, to use only open source information for obvious reasons.

The database has 46 variables and covers about 114 countries from 1945 through 1992 or sometimes 2000. These countries have a relatively large economy to be considered. We also include some 23 other countries which have engaged in proliferation. We have also defined different levels of proliferation, distinguishing between countries that have shown an "interest in exploring" from those who have "actively pursued" and those that have acquired at least one functional nuclear weapon.

There are a number of different modeling approaches that can be used, including an event-history approach and a multinomial-logit approach.[2] We have drawn up a list of variables that help explain what happens when nations proliferate. These include the factors that affect the behavior and decisions of proliferators at various levels outlined above. What kinds of insights can we get from the model?

Let's start with what the model predicts about states that have an extensive investment in civilian nuclear power. Matt Fuhrmann, currently at the University of South Carolina,[3] suggests that civilian nuclear power encourages an interest in weapons development. But our work suggests that perhaps that is not the case. It suggests that the more electricity you generate from nuclear power, the less likely you are to engage in nuclear proliferation.

If we look at countries that were part of the original Atoms for Peace program, quite a few seem to have at least thought about developing a weapons capability. But various factors hindered them — we call these "inhibitors." What about civilian nuclear power? Was it an "inhibitor"? What we found is that it depends on how much of a country's electric generation flows from nuclear energy. If you pursue a civilian nuclear industry in a major way, it's going to be a significant inhibitor. How significant will vary, depending on a variety of other factors: for example, does the country develop its own reactors, employ offsite fuel fabrication capabilities, etc.; how involved is the country in the international community; and how much industrial production is there? One must also factor in the presence of traditional IAEA safeguards. But the more you rely on civilian power production, the more inhibited you are. This is actually common sense. When you have invested in the creation of an infrastructure suitable for the development of a civilian nuclear industry, you are going to want to go on reaping the benefit of that investment. You don't want to become an international pariah at that point. Hence my model predicts that a nation with a lot at stake in the commercial nuclear sector may become more cautious.

We can also approach the issue from a different direction and ask: What are the determinants of successful nuclear power capability development? In his Introduction, Steven Miller stated that there are some 60 countries that want to be successful in this area. What do you need for this to happen? Let's think first about the countries that were part of the original Atoms for Peace project, the grand vision that President Eisenhower offered to the world. If you are part of the Atoms for Peace group, you get monetary assistance from IAEA. The list includes countries like Argentina, Pakistan, Chile, and Sweden.

Many of them have investigated nuclear technology and have started some work by importing technology, by importing research directors, by training people, etc. Of these, many of them don't as yet have nuclear power. Some like India, Pakistan, and Israel have nuclear weapons. Over time, more countries joined the Atoms for Peace group. But the picture doesn't change much. Even though these states started work, the majority still don't have nuclear power. Others have some capacity but not much. [4] India has spent a lot of money on civilian nuclear power development. But given how much has been invested, the amount of electricity it is getting is very small. We could even conclude it's a failed program as far as commercial development is concerned. Pakistan has likewise failed—it has two reactors producing 1.9 percent of its electricity, a meager amount.

What were the things that helped countries develop nuclear power—we call them promoters—or conversely, what were the inhibitors? Promoters included a good size gross domestic product (GDP), a considerable industrial capacity, major power status, the existence of a nuclear weapons program, a democratic tradition, Nuclear Non-Proliferation Treaty (NPT) ratification, and the existence of IAEA safeguards. Having a nuclear weapons program also was a factor in helping these countries develop nuclear power.

The inhibitors of civilian nuclear development included diplomatic isolation, frequent military disputes, enduring rivalries, and frequent changes in economic openness. Interestingly, getting IAEA assistance was an inhibitor—which means that countries who got IAEA assistance weren't able to make a good return on the investment. Corruption is also important. But we are looking for the variables with information that will help us complete our data. We haven't as yet added economic sanctions but are in the process of so doing. We also plan to include something about leader psychology, the characteristics of government bureaucracy, and so on. Those are very important variables but are not included in our current database due to lack of readily available data.

We are also looking at developing countries that attempted the development of civilian nuclear power capability as newcomer states. Unlike countries with a lot of resources, these don't have much to work with. There are 20 to 30 countries which, when they were still developing nations, initiated major civilian nuclear energy projects. States like Argentina, Brazil, India, Iran, Iraq, Mexico, Pakistan, Philippines, South Korea, Taiwan, Turkey, and Yugoslavia are all in this category. They all took some steps in this direction. Not all of them, however, got very far. For example, the Philippines and Turkey both spent a lot of money in an effort to develop civilian nuclear power. They did not succeed, however.

If we analyze what contributed to the success or failure of newcomer countries, we find that the factors at work are not much different from those operating among the initial Atoms for Peace countries. In particular, economic openness is very important along with the commitment to nuclear nonproliferation. If you want to become a successful country and if you want to become successful in developing nuclear power, you want to be open to the world economy while remaining committed to nonproliferation. But if you do something that awakens suspicion and you have a sanction imposed on you, and if you have a limited openness, that will hurt you. This is a good lesson to learn.

THE WATER-ENERGY-SECURITY NEXUS

Carey King

I'm going to talk about the work that I've done—most of it in collaboration with Michael Webber. We're engineers and geoscientists gathered to create technical information and phrase it in such a way that it can be used by policymakers. We chose to focus on the energy-water nexus and future changes, as this seemed to be one area that needs a little more thinking.

When we look at the water-energy nexus, we see it essentially as a two-way street. On one side, you need to consider the water requirements for producing energy, whether that's conventional fuels or new kinds of alternative fuels. You need to look at how constraints in water supply or concerns about water quality affect your ability to pursue energy production or conversion. On the other side, of course, you must ask how energy relates to the way one provides for water or for pure water for the public water supply. Furthermore, you must consider the restrictions on energy that can inhibit the ability to provide fresh water.

Earlier in the present work, we discussed disposable income in its relation to cost of energy. Water probably comes before energy. For developing countries or countries that aren't fully along the industrial chain, the provision of water is a major time-consuming activity that prevents them from pursuing other kinds of economic activities. So water is first on the list of things they need. If you can provide a supply of fresh water and the energy to distribute it—you can think of energy here as a type of infrastructure. Once it is securely established, only then will you have the time and the wherewithal to pursue other economic activities.

There are also water energy implications for places with marginal water supplies that have an advanced infrastructure. For example, we spoke earlier of the need for desalination in the Middle East. This need might be a reason for them to go for nuclear power. If people living in this region can desalinate water with nuclear power instead of using oil, they can sell oil in the market. Even before there were many oil wells in Saudi Arabia, there were likely desalination plants. These provided water for the workers, oil field personnel, and nearby cities. In that sense, the order of things was to provide fresh water first, then provide oil.

In the United States, the amount of energy spent on provisioning water, distributing water, treating waste water, and so on, is generally relatively low, depending on your perspective (about 3 percent of total consumption). But it can be higher in some places across the country. California is the poster child for the extreme case. Reports based on California's energy use show that 19 percent of all electrical and natural gas-based energy sources are associated with provisioning water in one form or another.

A lot of this expense comes from pumping water from the San Joaquin Valley over the Tehachapi Mountains to Southern California and the Los Angeles basin. Pumping water over this mountain takes a lot of energy, the irony being, of course, that Californians generate approximately 20 percent of their electricity from hydropower. Amazingly, the amount of energy they get from hydropower is the same amount of overall energy in BTUs that they use to distribute water since where people want to live is not where the water is.

On the other side of the street, the production of energy has its own water requirements. There has been earlier discussion of shale gas as a new energy resource. Water is an issue here because of hydraulic fracturing. Extracting the gas this way uses a lot of water. In fact, the process calls for something in the range of a few millions of gallons per frack well. Some of this water—maybe 20 to 50 percent—comes back up during the fracturing process and the drilling process. This water can be retreated and reinjected into a new well, or it can be treated at some sort of surface facility to remove contamination and discharged in the environment. Or it can be reinjected into a hazardous disposal well. Depending upon the geology, you have different kinds of options of what to do with this used water. Millions of gallons per well sounds like a lot of water. However, you get a lot more natural gas out of these wells for the water quantity, in terms of water consumed per BTU produced, than from many alternative and marginal energy options.

The fact is that, relatively speaking, hydraulic fracturing doesn't consume a lot of water, given how much energy you derive. But it sounds like a lot to the people locally. From a security standpoint, it's a particularly big concern to the people who live near the wells. After all, they're taking the risk of their own well water being contaminated. Water quality concerns are really the issues with hydraulic fracturing. This is because of chemicals included within the fracturing fluids themselves, but more so due to the water that flows back up the well during the fracturing process. The water thus produced is very saline and has other minerals and metals that exist in the deep subsurface. If something happens to a fresh water supply from a quality perspective, if the well is not sealed property, or if there is some kind of technical misstep, the local community members are the ones that may potentially suffer. It won't be natural gas recipients living hundreds of miles away.

From a quantity perspective, water has generally not been much of an issue so far in shale extraction in the Marcellus area[5] in eastern North America or in the Barnett Shale[6] areas of Texas. Water is relatively abundant in these places. The amount of water taken from the Northern Trinity/Woodbine Aquifer for fracking is less than 5 percent of total withdrawals.

Today in the Barnett Shale region, less than 5 percent of all water consumption from aquifers goes to hydraulic fracturing. It's not dominating the scale of use of people watering their lawns and doing all kinds of other normal things that people do on a daily basis. Yet it is 20 to 40 thousand acre-feet of water per year that was not being used before.

Compare that to the Eagle Ford Shale region in South Texas: here water is much scarcer and people depend more on and are affected by ground water limitations. People in San Antonio and areas like it are extremely vigilant about protecting their aquifer.[7] It is their source of water and livelihood. When you look at a region like this, you get a different perspective on the problem. Here, when you use a few million gallons of water per well and consider the number of wells that people are drilling, the hydraulic fracturing process can start to have a significantly adverse impact on the local water supply.

Geography matters in terms of this particular issue. Injecting hazardous waste, including low-level, into disposal wells is relatively prevalent in Texas, where the geology permits it. In Marcellus, the geology is not as amenable to deep injections of this sort. So there the people are concerned about how to deal with the wastewater from fracking. They get into issues like whether they should haul the water away somewhere to treat or inject it, or whether to treat it at local waste water facilities. They are concerned over the presence in the water of radioactive elements and other substances that originate in the shale formation. Their waste water treatment facilities are not necessarily geared to take that out of the water.

This brings me to carbon dioxide (CO_2) sequestration. Since the Environmental Protection Agency (EPA) has been making the rules in regard to this, and since the EPA has been charged with protecting underground sources of drinking water, a lot of the rules are associated with whether or not drinking water will be affected by carbon dioxide sequestration. Since carbon sequestration is not occurring on a large scale at the moment, the EPA is trying to preempt such problems by anticipating issues that are associated with potential ground water impacts from carbon dioxide sequestration. Geologists are quite confident that they can do this. In general, geologists feel they know what they are doing.

Drinking-quality ground water typically exists at levels shallower than 500 feet. Thus there has to be a fairly serious problem for underground sources of drinking water for it to be impacted from CO2 injection at several thousands of feet. This subject has been in the news lately. Take, for instance, the Weyburn Enhanced Oil Recovery Field in Canada, where carbon dioxide is injected for enhanced oil recovery. A nearby landowner has discovered some sort of leaking frothing emission from a hole that was dug. He paid a consultant to inspect the site.

The consultant concluded that CO2 was indeed coming from the nearby enhanced oil recovery operation. He did some measurements and figured he had identified the problem. But the Gulf Coast Carbon Center had a more cautious response. Their representative explained to me that the consultant essentially measured a factor that cannot determine whether injected CO2 had leaked to the surface. They really can't tell whether the CO2 is man-caused, in which case it would be coming from the coal gasification facility in North Dakota, or whether any excess CO2 is naturally occurring.

In short, methods more sophisticated than simply measuring CO2 concentrations are required to determine whether or not injected carbon dioxide has had an impact on ground water. One has to measure other chemical constituents in collected gas samples. It's quite difficult to tell what has happened far beneath the surface.

Turning to a different kind of alternative energy, biofuels obviously have a large water impact. Why is that? Because biofuels production is agriculture. To grow something, you must have water. Biofuels production, however, may or may not lead to an increase in water consumption, depending on circumstances. You may simply end up shifting water use from one sector of the economy, say, the food sector to the transportation or energy sector.

Let's consider this proposition. We have already discussed how you compare various transportation energy sources in terms of the water requirements — electricity, fuel cells, natural gas, unconventional fossil fuels, conventional fossil fuels, biofuels, and so on. These all have different units of measurement, a kilowatt hour, a gallon of water, a cubic foot of natural gas, etc. In thinking about how you would compare these to explain the impact of a shift to biofuels, it is useful to visualize the amount of water required to underwrite one mile of travel down the road using these different kinds of fuel. It turns out that a vehicle running on conventional petroleum requires somewhere around 0.1 gallons of water per mile; one running on natural gas is lower than that or about equal, even considering hydraulic fracturing. A corn ethanol E85 vehicle driving on irrigated corn ethanol uses in the range of 20 or 30 gallons of water per mile. One that runs on nonirrigated biofuels uses something like 0.3, 0.4, which is roughly the same figure you get when running on gasoline derived from oil sands. So unconventional fuels and nonirrigated biofuels look similar from a direct water consumption standpoint, but irrigated corn ethanol is a huge water guzzler.

Of course, from a water planning standpoint, you also care about precipitation, the total amount of water in a given basin, and what all the requirements for that water are. Thinking about biofuels is not necessarily any different than thinking about water resource management planning. You have to make a choice: "Do I want to allocate water for transportation fuels and feedstocks or for human food crops?"

To gain further understanding, you can look to the concepts associated with the virtual water trade[8] of products around the world. If you live in an area that is poor in water resources, you essentially have to earn money to buy products that cannot be grown locally because of the arid conditions, and you have to import agricultural products. Since dry regions have to spend money to import water because they can't grow crops, wetter regions grow crops and export these (and thus water) to the rest of the world.

If you apply this concept to the cultivation of biofuels, you come up with some interesting results. Is the United States not exporting virtual water, i.e., water-provisioned goods around the world? From a Brazilian standpoint, there's more water embedded in the ethanol associated from sugarcane than in our corn. However, the Brazilians have the water resources since it rains appropriately where they currently generate the sugarcane in the central southeast in the State of Sao Paulo. As they expand, they've got agricultural and ecological zones that will require some moderate amounts of irrigation at certain times of the year. They will then have to think about water provisioning and irrigation a little bit and the infrastructure associated with that.

We have resources that haven't been developed, such as oil shale. However, in regions like the upper Colorado River basin, water is a potential concern given the requirement of a few gallons of water for every gallon of extracted oil. This is where the water-energy nexus gets interesting in terms of local water impacts and global energy impacts, and where these push and pull each other.

From an electricity perspective, water is obviously important for hydropower generation. If you look at the capacity factors of the U.S. hydropower generation fleet over the last 30 years, you'll find that it has steadily declined. This decline is associated with the rise in all kinds of demands for water besides electricity provision, including irrigation and recreation. The planning associated with running a hydropower facility, especially given climate change and runoffs that come earlier in the year, is challenging, especially in the west.

The other main concern for us is power plant cooling. The main drain on water associated with electricity is such cooling. The percentage of water withdrawn for power plant cooling is essentially equal to the percentage that is withdrawn for irrigation. You may say, "Oh that sounds like a lot of water." But then you may ask how much water is actually *consumed*? It turns out to be 3 or 4 percent of the total nation's water consumption. That is how much is consumed at power plants for running cooling towers and cooling systems associated with thermal power generation, whether it's nuclear power facilities, coal, or natural gas.

It starts to get really interesting when you look at water rights in times of drought. A few years ago, the Southeast was hit by droughts. The nuclear power plants in Georgia were cutting back on electricity production because they didn't have an adequate or reliable enough supply of water to cope with the thermal management of the rivers. This is the process by which one takes cooling water from the river and then discharges it back to the river, while keeping it below a certain temperature.

From the perspective of those providing electricity, security concerns arise over maintaining the right sorts of temperatures to protect the environment and over whether or not there is an adequate water supply to do the cooling. For the most part, power plants have priority of water rights. For example, in Texas, if a drought comes along and water is scarce and people start cutting back on water supplies, most of the power plants are close enough to the front of the line, that they don't have to cut back on power generation until water flows get quite low. But from an environmental perspective, they may have to. Thus the drought scenario remains essentially a concern for electricity power production.

For that matter, concentrated solar power also raises water concerns. It's not just fossil-fueled thermal power plants. Cooling is a big issue with concentrated solar power in the desert. Obviously, it's very dry there, and water is tight. You can go to dry cooling, which raises the issue of dry cooling versus wet cooling.

NUCLEAR ENERGY AND THE MILITARY

James Bartis

We shall discuss a few issues broached in earlier chapters. With regard to the so-called nuclear renaissance that is taking place in the world, the question arising is, Will the United States be part of this? A few years ago, almost anyone who worked in energy policy would have said, "Absolutely," if for no other reason than that we have to deal with the problem of greenhouse gasses. But over the last 3 years, we have tripled our reserves of natural gas. It now appears that we have abundant natural gas resources and that these are obtainable at very low prices. It doesn't cost much to build efficient natural gas combined cycle power plants. So we have a choice as to what to do about our heavy dependence on coal for electric power. When we use natural gas instead of coal to make power, the greenhouse gas emissions are 50 percent lower. Thanks to the commercial development of natural gas from shale formations, the projected prices of natural gas are such that it looks like it will be competitive with coal for power generation.

So where does that leave nuclear energy in the United States? Nuclear energy doesn't produce any greenhouse gas emissions. On the other hand, the best estimates we have for nuclear power plants are that they're fairly expensive, especially if the alternative is a highly efficient combined-cycle natural gas plant. The recent MIT study estimated a range of $2,000 to $4,000 per kilowatt of capacity for a brand new nuclear plant. The U.S. Government has been trying to promote nuclear power and has offered loan guarantees and other incentives. Yet the costs that we're seeing are in the $6,000.00 per kilowatt range. If that's right, these estimates place nuclear power costs well above the competition. So it's questionable whether the United States is going to be able to participate in this renaissance. That's unfortunate because we have a lot to offer especially when we consider the potential for proliferation of nuclear weapons. It would be desirable for the United States to have a firmer standing in this renaissance.

Along with a few others at RAND, I recently completed a study on the nuclear fuel cycle. We looked at the back end, which means looking at how you manage spent waste from commercial reactors. There's a security issue here because some countries are talking about reprocessing this waste. We found that the kind of reprocessing technology that's available today really doesn't offer much benefit, if any, and it looks very expensive. They may have security reasons for reprocessing, but we don't see reprocessing as making economic sense.

Reprocessing doesn't reduce appreciably the amount of waste that's generated. It forces the country to develop a very large site that needs to be protected, that could be very polluted, and that could be expensive to restore. There are safe methods that we can use to manage spent nuclear fuel. Right now it's being stored at nuclear power plants. Our analysis says that dry-cask on-site storage is safe. We wouldn't want to store it there forever, but for the next 50 or a 100 years this is certainly feasible, and it doesn't cost much. While it's on the site, it's cooling down, which makes it a lot easier to ultimately dispose of it. Another option is to have centralized above-ground storage. For example, it could be cooled down at the front door of Yucca Mountain. There's certainly nothing wrong technically with going forward with the geological repository although, politically, we've made some big mistakes, and the government has not treated its citizens very well in the way it went about licensing Yucca Mountain.

Finally I'd like to turn to the use of nuclear power at military bases. A few years ago a Defense Science Board headed by James Schlesinger looked at energy, and one of its conclusions was that U.S. military installations were vulnerable to a loss of electric power. One reason they're vulnerable is that our grid overall is more vulnerable. Because of deregulation and other concerns, less investment is being directed at the reliability of electric power. The recommended solution was to look at renewable options, solar options for military bases. Since then, there's been a lot of movement in the U.S. military toward solar energy. A major motivation has been that this is a more secure source of power, but it's very expensive. I don't see much current effort in the military on nonrenewable options. If there is a security concern with power supply, what is the full range of options? How does the military work with the local utility to make sure that it has reliable power when it needs it and more secure power especially with regard to terrorist threats or natural disasters?

The nuclear development community in the United States has come up with the concept of a small nuclear reactor. In their view, it could be similar in size to, or smaller than, those that are in our nuclear submarines. The idea centers on developing a modular reactor that could be built in a factory, delivered, and assembled on site, all on an economical scale. This would be a constructive way to secure our military installations. When we deploy, we could even take a few of the small devices with us so we would have a power supply.

My own view of this concept is that we simply do not know whether it is at all reasonable. When we can't even price a conventional nuclear reactor within a factor of 2, I am skeptical whether we can price one of these brand new designs within a factor of 4 or 5. To me, it's an interesting option, the idea of putting a miniature power plant on a base, but at this stage, while it is probably worth further study, it is premature to reach any strong conclusions regarding its operational viability.

ENERGY, ENVIRONMENT, AND SECURITY

Alex Roland, James Bartis, Carey King, Steven Miller, and Man-Sung Yim

During the presentations, there was very little mention of the environment. I want to ask all our panelists whether they think environmental issues are security issues. What is the relationship between energy and security? Should we be thinking about environment as well? Do different energy choices have environmental impacts and do those environmental impacts have security consequences?

Carey King: The answer is yes. There are issues relating to the extraction of natural gas, environmental concerns about water contamination, and water treatment. We need to find ways to deal with that properly. But I think it could be handled. You need to plan it and put safeguards in place — many are already in existence.

James Bartis: In the defense community, this issue centers primarily on the destabilization of regions abroad as a result of climate change. There is quite a change taking place. What does it mean to places like Bangladesh that aren't stable and don't have the resources to mitigate environmental consequences and population shifts? Another issue is that energy requirements do consume water. To the extent that water resources are diverted for energy use, especially across borders, it is a security concern. That's primarily the way I see this played out.

Man-Sung Yim: Environmental issues are very important in terms of economics. People have tried to factor what they call externalities into cost. The cost of energy is more than it appears when we factor in the cost of the damage caused by, say, the release of carbon and other toxic materials produced. People have come up with a way to characterize the impact of certain types of energy use. That impact could include mainly health effects. Whenever we talk about environmental impact, we are generally talking about the effect on human health. When we talk about ecological impact, we are talking in terms of crop damage, materials damage, noise, air quality, acid rain, deforestation, and then maybe global warming. If you add these other costs to the cost of electricity generation, it will change the price of fossil fuels and renewables and nuclear power.

Alex Roland: Is nuclear waste an environmental issue?

Man-Sung Yim: Yes, of course it is an environmental issue. There are two communities of thought here. We have a technical community responsible for nuclear waste management. This community thinks the nuclear waste problem is solved; they think we have the technology to deal with it. If you talk to the other community, the problem will never be solved. As they see it, nuclear waste is a long-term problem. There are perhaps 1,000 toxic products in nuclear fuel, spent fuel. The vast majority of them are short-lived. Only 10 of the radionuclides have greater than 10 years half life—a very small number. But those few nuclides have a very long half life. Iodine 129 lasts 17 million years. So that's going to stick around forever.

James Bartis: And it's water soluble.

Man-Sung Yim: It could be water soluble depending on the chemistry of the system. Anyway, a couple of those radionuclides eventually are going to come and get you. The way the current regulatory system is actually written, you have to demonstrate that the waste will be safely disposed of for the next 1 million years, and that less than 100 milligrams will come from these radionuclides per year. If you take one CAT scan, you get 1,000 milligrams. If you take one X-ray, you get about 10 milligrams. Everybody in this room gets about 600 milligrams per year by just living in this country, so 100 milligrams is part of the natural background. But based on the regulatory system, you have to demonstrate that this nuclear waste disposer should not give you more than 100 milligrams per year for the next 1 million years. Indeed, for the first 10,000 years, you have to keep it to 15 milligrams. So suppose I come up with a system to dispose of the waste and try to demonstrate that these enormous multiple barriers work, I still have to face the challenge of predicting how they will behave, and be able to say that this system is going to work for the next million years!

James Bartis: The Department of Energy (DoE) was dishonest and disingenuous to the public, and this is part of the hubris of the nuclear development community.

Man-Sung Yim: Yes, actually there are a lot of challenges. There are a lot of challenges in the political area when it comes to nuclear waste. It is a very long-term problem, and you need a long-term stable policy. We don't have that. Policy changes whenever we have a change in the administration, and our administration has a bad record and little credibility in that regard.

Alex Roland: Is it a security problem or a political problem?

Man-Sung Yim: Both.

Steven Miller: I'd like to comment on the intersection of security and environment in the nuclear realm as it affects my little orbit. In a number of places, global climate change is one of the driving factors in pursuing nuclear power, not as an adequate response to global climate change, but as one of the portfolio of responses. Particularly in Europe where there's great environmental concern, countries like Sweden and Germany have decided to abandon nuclear power and are now reversing course. But others, like Finland, are investing more heavily in nuclear power. Insofar as you think the spread of nuclear power raises some security implications, there's a mixed pattern.

Another issue is the presence of nuclear assets in the context of conflict. We have to worry about nuclear facilities as targets in war or civil conflict. Nuclear technology and power plants these days are designed to be very safe; we have zillions of hours of operation with little or no significant incident. But what of a man-made Chernobyl? The human element is the weak link in all of these systems; moreover, an intentional effort outside a system to contaminate a site is possible.

During the Iran-Iraq war, the Iraqis repeatedly attacked Osirak. It was not yet operational with fuel, so that the implications were circumscribed. It is not entirely clear whether Iraq would have behaved differently if the facility had been operating. But there is some potential for deliberate targeting of a nuclear power plant.

Of course, the proximate nightmare that creeps into our current discussion is nuclear assets in the context of significant internal instability. So imagine that Maummar Gaddafi had not given up his nuclear weapons program. What would have happened to any highly enriched uranium in Libya? Or imagine everybody's favorite nightmare scenario, Pakistan, which contains 100 or more nuclear weapons and a pretty extensive nuclear infrastructure, along with the highest density per capita of al-Qaeda and Taliban supporters of perhaps any country.

COMPARATIVE SECURITY OF DIFFERENT TECHNOLOGIES

Alex Roland, James Bartis, Carey King, Steven Miller, and Man-Sung Yim

Of all the technologies we have looked at so far, are there any that are more inherently secure than others? At the other end of the spectrum, are there some that are so inherently destabilizing that some portion of our national and international decisionmaking should be based on choosing energy technologies that promote security and avoid choosing technologies that endanger the world. Some people think we fought the Iraq war because of oil, and some people fear that in the future we may face wars over water or because of pollution spilling across international borders. Are there energy technologies that are inherently more secure than others, and, if so, should we be leaning toward them?

James Bartis: Certain designs deter investment. Right now, if you start from scratch, you have to make a fairly large investment to get nuclear grade fuel. There are certain designs that make it very much less expensive.

Alex Roland: Assuming that most of the world's decisions about energy in the foreseeable future are going to be made on an economic basis, is there anything that the government can do at the margins to shape the way in which those economic decisions will present themselves? That is, should governments be worried about the security implications of energy choices, and if they are, what do they do about it? Is any one nation going to make the choice to make it a safer world and let everyone else get an economic advantage? Or do we have to get international agreement on this? As an example, I present the nonproliferation agreement where all nations agree that we will favor some paths of energy development as opposed to others which seem to be more dangerous.

Steven Miller: The first point I would make is that the market is a very powerful force in this sector as in many others. As our colleague showed, many states didn't pursue nuclear power even though they toyed with the idea. One of the more persuasive explanations is that, for much of the preceding 4 decades, nuclear power was not economically attractive. In my remarks, I touched on the Chernobyl effect. This definitely produced a backlash against nuclear power, but there's an argument floating around that basically says nuclear power was going to taper off anyway because the economics of energy were changing, and in an era of cheap fossil fuel and cheap electricity, nuclear power wasn't very attractive. Over the past couple of decades, you couldn't get an American utility CEO to invest a penny in a nuclear power plant. The same is true around the world.

The market remains a very powerful force except in those places where it's not wholly determinative. Where countries have made a national decision to go down the nuclear path because they need energy security or want national status, or because of some combination of self-interested factors, they have done so independently of calculations showing that this path was uneconomic or not the most optimal allocation of their resources.

France gets 75 percent of its electricity from nuclear power. The Japanese have made a massive nuclear investment, including exotic technologies, where they've invested trillions of yen without a penny of return in terms of generated electricity. The Koreans are now doing similarly. As to the Chinese, their long-term vision is grandiose. I was just in Beijing recently. They have a pipedream plan—not approved or funded—to produce 150-200 reactors in the next 50 or so years. In that wonderful Chinese way, they have a slogan, "100 in 100," meaning 100 percent replacement of fossil fuel with nuclear in 100 years. The way they propose to do that is by using breeder reactors. This plutonium recycling would bring them way beyond the light water reactor era that we're living in today. If you were to get a government to commit to that kind of path, then some of these market calculations would begin to be less of an impediment.

Alex Roland: The changing rationale was what? Security, energy independence?

Steven Miller: The present Chinese program will result in at most about 4 percent of China's electricity being generated by nuclear power. That's the part that's most likely to be real because the further out you get in somebody's vision timeline, the less likely it becomes real. The hoped-for program is a 100-year pipedream. It's just suggestive of Chinese enthusiasm.

James Bartis: There's another factor. China is using a tremendous amount of coal, but their reserves are not that large.

Audience: They're the third largest in the world.

James Bartis: I believe there are about 50 billion tons of coal in their reserves. They have about 25 to 30 years worth of supplies. There may be more coal undiscovered, but given how fast they are putting in new coal plants in China, they are doing the math and saying to themselves, "We've got to worry."

Audience: The bigger problem for China right now is the lack of an effective electric grid distribution. They can't even distribute what they're making off the Three Gorges Dam Complex, let alone all this new electricity — that itself is a pipedream.

Man-Sung Yim: An additional factor we need to consider as part of the security issue is indeed electric grid security. As an electricity-generating country, you have to look at the changes in demand. Some plants run all the time, like base load plants.[9] Some plants fluctuate in output, running intermittently. Some plants run only at times of peak demand.[10] These patterns have to be a part of this discussion. Going back to the question about security, that is part of the reason that President Obama wanted to talk about nuclear power in the State of the Union address. He wants to reduce our foreign oil import dependency, and you cannot make a dent in this without resorting to nuclear power. That is why a lot of countries are interested in nuclear power.

James Bartis: An earlier panelist made a very good point — we don't use oil to make electricity in this country.

Man-Sung Yim: Actually, at this point it's natural gas mostly. But to get to the question, economics is the driving force for this country, unfortunately. No matter what the government decides — and right now it's fully supportive of the nuclear renaissance — economic considerations will be determinative. There are other countries — for example, South Korea, Japan, China, and Russia — where we're going to see the renaissance. That's not really a threat because they already have the capability. We're not going to see many newcomer nuclear countries. It's very difficult to develop civilian nuclear power capability.

QUESTIONS AND ANSWERS

Alex Roland: Now for audience questions and comments.

Q: My question is for all the panelists. Underlying a lot of this discussion is the question of intent. When we're talking about nonproliferation and the expansion of nuclear energy in places where it hasn't existed previously, there are some places where we're going to care if they want nuclear energy and some places where we don't care. If Canada wants to have a massive expansion in the number of nuclear power plants they have, we're not going to care. But there are other places, particularly in the Middle East, where the fuel cycle is going to be a major concern. Dr. Yim's presentation gets at how we try to understand national intent. What makes countries decide to pursue nuclear weapons? Does it have any relationship with the development of civilian power? What is the U.S. Government's view on how these decisions are being made? What calculations do we make in determining countries' intent? Will we not mind if some countries develop fuel cycle facilities? Or are we going to say across the board, "No, we're not going to help with technical assistance if you want fuel cycle facilities in your country?" Earlier, it sounded as if the countries that will actually succeed in pursuing nuclear power and build those 65 new facilities are those countries that we don't see as a threat.

Steven Miller: Actually, under the guise of the universal nonproliferation policy, we've always had two nonproliferation policies, one toward hostile proliferators and one toward the friendly. Even though we would prefer that no additional states have nuclear weapons, when friendly states have done it or sought to do it, we've reacted with reasonable tolerance. It's the hostile cases that worry us.

For example, we were worried by the Soviet Union, China, Iran, North Korea, at one point Libya, and at one point Iraq. We always react negatively whenever one of these cases comes up. So you're right that we don't have equal reactions to every case. Over the last decade, we've objected vociferously to Iranian efforts. But the Brazilians, simultaneously, have developed a commercial enrichment facility, and they weren't subjected to any of the scrutiny or ostracism that the Iranians were. They didn't get brought before the United Nations (UN) Security Council. They weren't sanctioned. The United States wasn't enthusiastic about the Brazilian development, but we didn't actively resist it.

That raises a question of whether you can have a universal nonproliferation regime with exceptions that exempt our friends from various restrictions. We did the same thing with India in the U.S.-India deal. We had strategic interests in pursuing a relationship with the Indians, and so we exempted them from various strictures that we had worked very hard to establish as norms of the nonproliferation regime. We are then accused of hypocrisy. We then often try to persuade ourselves, "Well, this is going be a unique case." Then, lo and behold, not long after the United States-India deal, the Pakistanis and Chinese wanted to do something similar. Now there are noises bubbling around in the background as to why Israel can't have some kind of special arrangements, etc.

These are exactly the kind of tricky issues of managing the nonproliferation regime in circumstances where the fact of hostility or friendliness has a huge impact on how we respond to things. I offer you one further example. In my opinion, Japan has everything that would permit it to become a nuclear weapons state. It has enrichment, it has separated plutonium in large quantities at Rokkasho-mura. You can visit the facility. It has a reprocessing facility. It has high levels of nuclear expertise. It has delivery systems from the United States that are dual use. It has everything to be a nuclear weapons state other than the decision. It can be regarded as a deeply dual nuclear weapon state. This doesn't bother us terribly much because the Japanese are close allies and we don't doubt their intentions. It does at least modestly trouble some of their less friendly neighbors, but if Iran were in the circumstance today that Japan is, we would be extremely agitated because our relations are hostile and we do doubt their intentions.

James Bartis: There is another a lesson here. Iran looked at the situation with North Korea, and they looked at what we did to Iraq. Accordingly, they probably made a national security decision to develop nuclear power. From their perspective, that may make a lot of sense. As has been mentioned, a lot of countries don't base their decisions solely on economics. But when we look at it from an economic perspective, we can understand why Kuwait, for example, might want to build a nuclear power plant. They're using oil today to make electricity, which means it's very expensive. Moreover, they don't have much natural gas. So developing nuclear power may make some sense there. Iran has the second largest natural gas reserves in the world. It's almost free there. It's inconceivable from an economic perspective that the Iranians need to use nuclear energy to make electricity. Moreover, Iran is not a state that's very well developed compared to many other countries, and they could certainly use the money they are using to build nuclear power plants in more productive areas. But they're doing it totally for a national security reason, and I think we're part of the reason why they're doing it.

Man-Sung Yim: Actually, if we look at history, many countries, for example Japan, South Korea, and Taiwan, initially developed civilian nuclear power because of their interest in weapons. That's just a fact. They were looking at both civilian capability and weapons capability. After the bombing of Hiroshima and Nagasaki, the Japanese came up with a political agenda. They concluded they eventually had to develop a nuclear capability. They came up with a 235 million yen budget; 235 is a symbolic number, and they wanted to develop the capability.

Then South Korea was struck by the same idea. They actually formed a plan to develop both military and civilian capability. Taiwan did the same thing. Today, the way the nuclear energy infrastructures have developed means that you cannot just go off and do something (like build weapons) and have nobody know you've done it. While they do have the technical capability to build weapons, they have a very long and robust institutional investment in sole-civil use, so that's why we don't see them proliferating.

Other countries like Taiwan and South Korea also have a similar long institutional investment so they are not likely to turn their civilian energy program into a weapons program. Then there are countries like Vietnam and Iran. I believe the Vietnamese leadership may be interested in civilian nuclear power because they have some interest in nuclear weapons as well. They have good reason to do so: they have China and India as neighbors. Vietnam also has the capabilities: they are a very smart people. They're a capable country.

But even if you start out with intending to develop nuclear weapons at some point, the international regime is not going to help you. We have this additional protocol, the UN Security Council Resolution 1540, and the Proliferation Security Initiatives (PSIs).[11] All this involvement in the nonproliferation regime is going to hurt you, so it's becoming very difficult. That's part of the reason why many countries who pursued both civilian and weapons programs had to give up the weapons program. It would actually have a very negative effect on the civilian side. They have to make a very calculated decision. Most of the countries have chosen not to pursue nuclear weapons to avoid a potential backlash.

Q: This question is mainly addressed to Dr. Bartis. What does the U.S. Government need to do to finalize a solution to the waste storage issue? Yucca Mountain fell apart even though it seemed as if that deal was totally implemented. If we are going to find a solution, what must happen?
James Bartis: Early on, Congress authorized the DoE to look at numerous sites for a waste facility, and then a few years later, Congress itself decided on a location. They picked Yucca Mountain in Nevada. Nevada has a small congressional delegation, and at that time none held positions of great power. Little did they know that a senator from Nevada that was deeply opposed to the selection of Yucca Mountain would become the Senate Majority Leader. The Yucca Mountain deal also failed because of the fairly roughshod way that the DoE went forward with the project.

Yucca Mountain may or may not be reconsidered as a site—it's certainly a viable one. Our research at RAND finds that there is no requirement to find a geological repository site within the next 25 years. So it may be appropriate for the United States to restart the search for an appropriate site, take its time, treat the public appropriately, evaluate multiple sites, and determine which site is most viable. There are communities in the United States that may be very interested in the economic benefits of having a nuclear waste disposal site in their vicinity. That certainly was the case in New Mexico where there wasn't unified opposition to the geologic disposal site for defense-related nuclear waste materials.

Audience: Are you saying Yucca Mountain is not a good site for political reasons?

James Bartis: Absolutely. Certain claims were made on behalf of Yucca Mountain and those claims over time turned out to be incorrect. As a result, the DoE recommended what we call "engineering solutions" so that, in addition to the geology, there would be engineering barriers. A number of concerns were raised about some of these engineering barriers.

Basically, there was a breakdown of trust between the DoE/federal government and the State of Nevada, and it was a tremendous waste of taxpayer money. A lot of the blame here goes to the federal government for the way the DoE behaved. A solution to all the problems in Yucca Mountain was to simply keep the material above ground for at least 50 to 100 years. After 50 to 100 years, the radioactivity and the heat generated by that radioactivity goes down significantly. It's the heat generation that poses the problem underground. If you keep it above ground and the waste has time to cool, there is much less risk associated with the integrity of geological disposal.

Audience: Do you think the same political problem would exist if we moved all of the fuel to the front door of Yucca Mountain?

James Bartis: Even that is unlikely to work now because there has been a breakdown in trust. There needs to be a new procedure for site selection before you even start thinking about that option. I don't know whether the DoE has formulated an acceptable strategy for dealing with the public.

Audience: The issue is not the site. The issue is the fact that you have to move 80,000 tons of waste on either beltways or highways, and you need equipment that's going to be operated by human beings. All it takes is one derailment or one truck accident, and you could have a major problem. John McCain, for example, said that he supported putting the waste at Yucca Mountain. He was asked, "Would you agree to have it go through Phoenix?" "No, I would not." That's the problem.

Q: A comment. The United States is highly disturbed over the prospect of Iran developing what it calls a civilian nuclear program, and we know that's not what they intend to use it for. Yet, we warmly embrace the Sunni Arab countries, Saudi Arabia, Yemen, Egypt, etc., that want to go in the same direction. Thus when you think of civilian nuclear programs in the context of the Middle East, it's probably an oxymoron.

Steven Miller: Two points. States have a right, and in the context of civil nuclear power, the right is enshrined in Article 4 of the Nonproliferation Treaty, to make whatever choices they wish in the realm of nuclear technology. Everybody's heard of the phrase "the inalienable right to exploit nuclear technology for peaceful purposes." If you actually read the second paragraph of Article 4, it has a phrase stating that you're entitled to the fullest possible exchange of nuclear technology. So the language is expansive rather than limiting. Whether we like it or not, a lot of states are going to at least explore these possibilities.

In the case of Iran, we have lots of reasons for thinking that they have ulterior motives. But they've been struggling, almost completely unsuccessfully, to implement the program that was announced in April 1974 by the Shah and was worked out with the full connivance of the Nixon/Ford/Kissinger administrations. True, it produced a huge fight in the U.S. Government because the nonproliferators were worried about what the Iranians were really up to. But the gross anatomy of the current program of the Iranian government is essentially identical to the Shah's program: 20 large light water reactors, and the full fuel cycle, and so on. This is not an idea the mullahs came up with; this is an idea the Shah worked out with Henry Kissinger. So whatever else they might be up to, they do have an interest in civilian nuclear power.

The final point I would make is that Abu Dhabi is the spearhead of the nuclear newcomers. It's they who will be the first to put a nuclear power plant in the grid and actually generate electricity among the newcomers. They negotiated a 1-2-3 agreement with the United States in which they agreed to everything that we wanted them to agree to. They had made a national decision that they wanted to move as quickly as possible toward nuclear power, and so they understood that they needed to dispose of this 1-2-3 question as quickly as possible. Therefore, they judged it to be in their own self-interest to acquiesce in all of our preferences. They agreed not to enrich. They agreed to forsake their right to reprocessing. No fuel cycle. In Abu Dhabi (or the Emirates), they agreed to sign on and adhere to the additional protocol of all other safeguards arrangements that are currently the norm.

In the nonproliferation world this has been dubbed the Abu Dhabi model. If all the nuclear newcomers will do this, the proliferation implications of nuclear power are much more circumscribed.

It turns out that lots of other states are really angry at Abu Dhabi for its course and refuse to do it themselves, including Vietnam which has been in similar discussions with the United States and has refused to forfeit its right to enrichment. The Egyptians are also in the same position. The Saudis have said they won't sign up to such an agreement. I was at a meeting in the Gulf this summer at which one Arab and Gulf state after another expressed regret that Abu Dhabi's lack of foresight in acquiescing to America's full spectrum of preferences.

James Bartis: If I look at Iran, so long as Iran allows unimpeded inspections—and right now we have no evidence that they're doing anything that's against any letter of the law—they will be compliant. Is that correct?

Steven Miller: Iran committed, over a very protracted period of time, a substantial number of safeguard violations. These violations fell entirely in the domain of reportage. They were required to report activities, provide information, etc., which they systematically failed to do over a 20-year period, resulting in a long list—bill of indictment against them by the International Atomic Energy Agency (IAEA). What got them in trouble was not that they were pursuing enrichment, which is a permitted activity under the treaty, but, rather that they failed to report that they were pursuing enrichment.

The way we trapped them was by using the safeguard violations to claim that they were delinquent in their NPT obligations. That gave us the right to refer the matter to the UN Security Council. We then issued a series of the UN Security Council resolutions under Chapter 7 which are, in principle, binding which instructed Iran to cease and desist its enrichment and plutonium-related activities immediately. Their failure to do that then puts them in a position of noncompliance with international law under the UN charter. So that was the legal mousetrap that we set for them. But under the Non-Proliferation Treaty (NPT) or its safeguards agreement, it has every right to both enrich and reprocess so long as it permits the required safeguards under Article 3 of the treaty.

One last point: Iran is the most heavily inspected party in the history of the system now, and it has been subjected to more short-notice inspections than all other states in the system put together. There's no evidence that the known and declared facilities are being misused for weapons purposes. So the fundamental issue is what covert actions or activities are underway. That, of course, is what the IAEA is never given access to.

Audience: What's the case on North Korea? Similar to Iran?

Steven Miller: North Korea belatedly signed a safeguard agreement when the inspectors came in. In 1991 the inspectors detected discrepancies in the material accountancy which revealed quite unambiguously that North Korea was diverting material. In 1992 the IAEA declared North Korea to be in a formal state of noncompliance with its obligations under the safeguard agreement and referred the case to the UN. At no moment since then has North Korea ever been restored to compliance. The UN Security Council delegated the issue to the United States which, in 1994, reached the agreed framework arrangement with the

North Koreans which froze their plutonium program for about a decade. But in late 2002 or early 2003, partly motivated by and partly under the cover of our run-up to war with Iraq, the North Koreans untagged and unsealed the tagged and sealed fuel rods that were under IAEA surveillance. They ripped out the close-circuit television cameras. They threw out the IAEA inspectors. They withdrew from the NPT. They restarted their reprocessing facility and in due course they detonated nuclear weapons.

The timing of North Korea's joining NPT coincides with the completion of their reprocessing capability and with South Korea's and North Korea's joint declaration of denuclearization of the peninsula. Thus, basically, whenever North Korea takes some liberty, it feels comfortable in giving up something. But there is always covert development behind it.

Q: To bring the discussion back to environmental issues, there's a growing body of literature linking security challenges with environmental change, whether it be climate change or water depletion, you name it. But if we look at where carbon is going to be coming from over the next few decades, it's mainly the developing world because coal is cheap, and its inefficient technology will be causing the vast majority of new carbon emissions globally. My question is, How do we deploy alternative technologies in the developing and poor world, which will be driving these security climate threats for the next decade or century?

James Bartis: Actually, it's not the poor world that's going forward with coal that much. If you look at where the action is going to be, it's China. It's a very limited set of countries that are going to be putting out large emissions, but China and India are two of the really less developed countries that are going to be doing so. That's partly because these countries have very large populations. However, even when you look at extensive development over the next 30 to 40 years, their carbon emissions still are not close to ours. What we need to do is set an example—we've got to take some steps in the direction of developing alternative energy.

Carey King: Alternative energy is going to be important as a tool for adaptation if some of the problems we expect with climate change materialize. It will help provide basic logistical capabilities and basic water provision. If you have a decent number of solar panels and a decent number of wind mills, you can desalinate some water. That will create resilience. Unless the grid gets shut down or broken because of floods or earthquake, alternative sources can provide the basic needs in difficult times. They can take a bad situation and prevent it from becoming really horrible.

Steven Miller: When it comes to mitigating climate change, it is with the big emitters that we can find traction. China's gross domestic output (GDO) per capita is nowhere close to ours, but where they've surpassed us is in terms of gross carbon emissions. The problem has to do with long-lived power-producing assets. Some of our colleagues say, only half tongue-in-cheek, that the Chinese are building nuclear power plants faster than humanly possibly, but they're building coal and natural gas fire plants even faster. Those facilities are going to be around for 30, 40, or 50 years. Thus the decisions the world actors are making today we'll be living with decades down the road. A lot of them are still running in the wrong direction.

James Bartis: The Chinese are making major investments in technologies that are energy efficient. They're making advancements. They're not going down the path that we're going down.

ENDNOTES - CHAPTER 3

1. See, for example, Dong-Joon Jo and Erik Gartzke, "The Determinants of Nuclear Weapons Proliferation," *Journal of Conflict Resolution*, Vol. 51, No. 1, February 2007, pp. 167-194, available from *jcr.sagepub.com/cgi/content/abstract/51/1/167*.

2. The logit model is also known as the logistics model. It is a logistic regression used to predict the probability of occurrence of an event by fitting data to a logistic function.

3. Fuhrman has since assumed a position at the University of Texas-Austin—he recently participated in a workshop at North Carolina State University on "Securing our Nuclear Future" and gave a seminar to the Triangle Institute for Security Studies.

4. The slide showed at the conference contained a list of current capacity as of February 1, 2010, in states to include: the United States, France, Japan, Russia, Germany, South Korea, and Ukraine.

5. The Marcellus Formation is a huge unit of marine sedimentary rock found in eastern North America. It extends throughout much of the Appalachian Basin. The shale contains largely untapped natural gas reserves. It is near to high-demand markets along the East Coast of the United States, which makes it an attractive target for energy development.

6. The Barnett Shale is a geological formation located in the Bend-Arch-Fort Worth Basin. Arguably, it has the largest producible reserves of any onshore natural gas field in the United States.

7. The Carrizo Willcox aquifer is a vital source of fresh groundwater for much of South Texas.

8. Virtual water is water used in the production of a good or service. It is also sometimes called embedded, embodied, or hidden water. The concept has been influential in discussions of water management. It helps explain why some nations export and import water.

9. A baseload plant or power station is an energy plant devoted to the production of baseload supply. This is the amount of power required to meet minimum demands based on reasonable expectations of customer requirements. Such plants (which typically include nuclear and coal-fired plants) produce energy at a constant rate, usually at low cost relative to other production facilities available to the system. Some renewable energy sources can provide baseload power.

10. Peaks or spikes in customer power demand are handled by smaller and more responsive types of power plants called peaking power plants, typically powered with gas turbines.

11. Resolution 1540 (under Chapter VII of the United Nations [UN] Charter) was adopted on April 28, 2004, by the UN Security Council. It obliged states to refrain from supporting by any means nonstate actors in developing, acquiring, manufacturing, possessing, transporting, transferring, or using nuclear, chemical, or biological weapons and their delivery systems. The Proliferation Security Initiative (PSI) is a global effort launched in May 2003. It aims to stop trafficking of weapons of mass destruction, their delivery systems, and related materials to and from states and nonstate actors of proliferation concern.

CHAPTER 4

ALTERNATIVE ENERGY FROM A SECURITY PERSPECTIVE

This panel was organized as a "conversation" between a moderator and panelists. Its purpose was to highlight the ways in which existing and future energy technologies affect security at the human, national, and collective levels. The focus was on alternatives like wind, solar, and biofuels, though attention was also paid to shale. The moderator was James Trainham. Panelists were Michael Roberts, David Dayton, and James Bartis.

HIDDEN COSTS OF ENERGY

James Trainham, James Bartis, David Dayton, Michael Roberts, and Daniel Weiss

Let's continue our discussion on the use of renewables and alternative energies for energy security.

James Trainham: As an engineer, I work with balances: charge balances, mass balances, momentum balances. I also work with cash balances because engineers do take economics into account. One of the things we talked about a lot today is the market. I'd like the panel to discuss the real cost of our existing energy infrastructure. A number of studies note the hidden costs of our existing energy supplies. For example, just recently a Harvard Medical School report stated that coal is costing healthcare in the United

States between $140 and $300 billion a year. That is not reflected in the pump price.

Likewise, some suggested today that "we could pull our military back." If you are trying to find out how many military installations the United States has in the Middle East and Africa, you'll find that the list covers a page, not counting what's in Afghanistan and Iraq. That also covers a page. There's a website that will show you where all these military installations are. So you have to ask, "What does that cost and who's paying for it?" It's obviously the American taxpayer who's paying for that to keep oil flowing.

Thus we have a lot of hidden cost, and we can assign that cost a number. Shouldn't we also be looking at the alternative energy costs? They are always being criticized on the grounds that they're being subsidized. Let's look at the hidden subsidies and see what it really costs to produce various energy sources. Let's look at the human costs and the societal costs we're paying for the existing infrastructure, and compare the total balance of cash and costs. That real cost should influence our decisions as we make our moves toward alternative energy.

Do we institute various taxes to influence the energy market? How might we address the total cost so that we actually make the right economic decisions instead of making economic decisions just because we have institutionalized our current energy sources?

Michael Roberts: I mostly do work on agricultural commodity prices, and I'm doing some work on the effects of ethanol on commodity prices worldwide. To answer your question, in general the economist solution to problems like this is to think in terms of externalities. An externality is some kind of effect from a market transaction that affects someone besides the people trading something on the market, say, oil. In other words, I buy my gasoline and drive my car. The gasoline is going to cause pollution that affects everyone, but I'm not paying for the price of pollution. So the economist's solution to this is to put a number on the external social cost and factor that into the price of gasoline.

Gregory Mankiw, a conservative senator who was a leading advisor to President Bush, and a leading economist at Harvard, has written the best-selling text book on the principles of economics.[1] He started something called the Pigou Club, so called after the Pigouvian tax. The idea behind this tax is that we should put a tax on these external costs.[2] So if we think that carbon emissions are causing global warming and that there are external costs related to that, we should put a price on it. We can tax these externalities, and then perhaps even lower income taxes or other sorts of taxes. We should tax the things that have social costs and thereby increase efficiency rather than reduce it. Even this fairly conservative economist highly recommends these kinds of taxes, and I think that's a rather standard economist answer to that problem.

Of course, there are other ways of doing it. Cap and trade is another way of putting a price on externalities, and we did that with sulfur dioxide markets. We introduced tradable permits for sulfur dioxide around 1995-96, and this effectively put a price on sulfur emissions. This mechanism very rapidly lowered the costs of reducing sulfur emissions. This approach is in contrast to command and control mechanisms where you order the electricity plant to employ scrubbers or other particular techniques. Instead of forcing their hand, you simply put a price on the offending pollutant and then let the markets work it out.

That's the standard economist answer to your general question, and I favor that sort of thing. It can be politically challenging to do, as we've seen time and again. But in cases where it has been done, it's been, for the most part, fairly successful.

James Bartis: When I look at just gasoline — not coal, for the moment, but just gasoline — I see four associated externalities that are not being accounted for. In other words, there are four additional costs that should be factored into the price of gasoline. I'll just list them here and try to provide some numbers.

In the first place — and this is the most egregious example — the gasoline tax doesn't pay for the highway infrastructure in any state of the United States. Instead, we subsidize drivers by taking money from the general funds of the states and the general fund of the federal government. The gas tax needs to be 25 to 50 cents higher per gallon to break even on the infrastructure requirement. Then there is the energy security problem that goes along with having a cartel at play. We at RAND and some others have looked at that externality and calculate it at about 25 to 50 cents a gallon. There's also a smaller externality called the supply risk. A recent paper by two reputable economists suggests that's about 15 cents a gallon. Then there is the greenhouse gas problem. We often hear proposals suggesting that there should be a charge for releasing carbon dioxide into the atmosphere. For each $10 charged for emitting a ton of carbon dioxide emissions, the price of gasoline would go up by roughly 20 cents per gallon. Finally, gasoline use has health costs. Even though we have much cleaner cars, there is still a particulate problem. I don't know what the size of that externality is. My very rough guess is that it is at least 10 or 15 cents per gallon. If you put all these externalities together, you ought to have a gasoline tax that's about $1 to $1.50 higher than it is today. In effect, we subsidize gasoline to that extent.

There are other subsidies directed at promoting domestic production of oil, and the administration is trying to address those right now. Our failure to effectively address the externalities associated with oil use is one of the key problems of our current national energy policy. If we had a gasoline tax $1.50 higher than the one we have now, we would be using less gasoline and we wouldn't drive such large cars. After all, the Europeans live very well. They use half the petroleum we use on a per-person basis. So there are benefits to increasing the gas tax to reflect externalities.

The reason we don't do this is that there's a lot of wealth associated with the current system. There's a lot of money to be had, and there's a lot of inertia that gets in the way of our doing anything that moves us toward a rational energy policy. Instead, we continue to rely on the hope that there's some magic research solution out there, that some new technology is going to come in and solve all our energy problems. We make our problem worse by subsidizing the wrong fuels. If we look at some of the renewable fuels being pushed by the federal government, we'll find that there are also significant externalities associated with them.

David Dayton: That was a very good segue into what I want to say. Unfortunately, for things like alternative fuels, you have to consider the entire value chain from the start—the feedstocks, biofuels feedstocks, all the way to transportation, marketing, and distribution to get the fuel to the pumps.

It's going to be hard to develop new technologies, at least in the current environment in which the infrastructure is geared to fossil fuels. Until we do, we won't be able to determine what those externality costs are. That's because we've got such an uphill battle to fight. How can alternatives compete with cost-competitive fuels like petroleum-derived distillates, especially given that these alternatives are inherently higher in cost to start with? There's really no economic incentive yet to adapt to new technologies.

There are some known externalities. In particular, there are issues connected with feedstock distribution. Biomass inherently is much less energy dense than all the other fossil fuels that are being consumed in the country.Therefore, you've got a huge logistics problem. Leaving aside farming practices for a second, you've got to move material from the farm or growing site to a processing plant. There are associated costs with that are more challenging than pipelining and having supertankers sail through the Straits of Hormuz filled with fairly cheap crude oil.

Then you've got to consider conversion technology options. Right now, the only biofuel that really is in the marketplace is corn ethanol, on the order of about 10 to 12 billion gallons per year. There's a smaller market for biodiesel from vegetable oils, about a billion gallons per year. But the scale of the fuel infrastructure, just for motor gasoline, is about 149 billion gallons per year. At 10 percent ethanol, we've already maxed out on the permissible fraction of ethanol in gasoline (i.e., the "blend wall"). We're looking at new policies to raise that fraction.[3] But even if new technologies were developed, let's say, for ethanol, which may or may not be a good fuel, there's no market for it. That's because there are not enough flex-fuel vehicles to use it nor a higher blend wall.[4]

I am an advocate of looking into advanced biofuels as a way to replace hydrocarbons. The energy balance for those fuels is even more challenging than it is for ethanol because you can't retain any of the oxygen that's in biomass in your fuel. You've got to get rid of all of it. That means you must go from 40 percent oxygen in your biomass down to 0 percent in your fuel. So you are going to be dealing with a 50 percent weight loss when you go from starting material to fuel. Fortunately, your energy density can be upwards of 50 to 80 percent, depending on your conversion technology.

One of the additional advantages of looking at advanced hydrocarbon replacements is that you get to use the existing infrastructure. I think that's one of the keys for developing these alternative technologies.[5] We've already got a trillion dollars of capital invested in refineries, pipelines, and marketing and distribution infrastructure for hydrocarbon fuels. It would be rather silly not to take advantage of this, at least in the near term. We have an opportunity here.

Daniel Weiss: I just want to add one thing to what James Bartis was saying, and that is that the National Academy of Sciences did try to put a dollar figure on the healthcare costs of combustion of fossil fuels. It found that the added cost is approximately 50 percent for oil and 50 percent for coal, give or take. The Academy came up with a figure of $120 billion a year in additional healthcare costs due to premature deaths, lost productivity, hospital visits, etc. That's a good starting point for trying to figure out how to integrate those costs.

James Bartis: I will address the issue of how far biomass or biofuels can go in addressing our national energy needs. It's important to understand where we are today as well as where we might be able to go in the future. Right now, we get almost a million barrels a day from ethanol. That's a significant amount of alternative fuel. The question is: How much farther can we go? To answer that question, we can look at the different options available.

One option—the nearest-term option—available to us is to use seed oils. At RAND we looked at using seed oils like soybean, camelina, or jatropha.[6] We asked, "How far can seed oils take us?" We found that the productivity of seed oil per acre is fairly low. It turns out that to get 200,000 barrels a day of oil from seeds—which is 1 percent of national oil/petroleum use—we'd have to use 10 percent of the cultivated land in the United States.

One of the companies that advocates for camelina-derived seed oil calculates that the maximum sustainable national production for that seed oil is about 45,000 barrels a day. That's fine if you're in that business, but from a national perspective, that's not going to solve any of our problems. The National Academy of Sciences and others have also looked at using biofuels where not just the seed but the whole plant can be used, as in the case of wood or grasses. Their estimates vary. Oak Ridge came up with a billion tons a year. When the National Academy looked at that figure, they knocked it down to 400 million tons a year. They did say that over time that figure could reach a half a billion tons a year. If we used all of that biomass to make liquid fuel, we could make about two million barrels per day. Now, that's an appreciable amount, on top of the million barrels per day already from ethanol.

Biomass also offers us some other opportunities. We can co-fire it with coal in power plants. Also on the horizon are some interesting technologies associated with algae and certain microbes. These organisms produce oil through photosynthesis.[7] There's a fairly strong consensus that algae-to-fuel approaches are still in the research stage. They are potentially important in that they free us up from dedicating massive amounts of our land to energy crop production, be it poplar trees or certain varieties of grasses. So the possibility is out there that, with these genetically modified microbes and algae, we could produce substantial amounts of liquid fuels.

Right now we can estimate the extent to which biomass can take care of our energy security. We use almost 20 million barrels a day of oil. Maybe we can meet 10 percent of our needs from alternatives. We already meet 5 percent by using ethanol. Maybe we can get another 5? Or even 10? Altogether, 10 or 15 percent of our petroleum consumption might be met with biofuels that are emerging now or are already available. In another 10 years, I might be telling you a very different story, if we're lucky and if our researchers are successful. But it's important to put that in context. Biofuels do not get us off oil, but they help. Having two million barrels a day of lower demand just from the United States will do something. Hopefully, other nations will also reduce their use of conventional petroleum. It all adds up. Reduce demand for Organization of the Petroleum Exporting Countries (OPEC) oil. by five or six million barrels a day worldwide, and it makes a difference in what the world oil price will be.

CLIMATE VS. ECONOMICS: SECURITY IMPLICATIONS OF ENERGY CHOICES

James Trainham, James Bartis, David Dayton, Michael Roberts, and Daniel Weiss

Arnold Schwarzenegger spoke at a recent energy conference, where he made two basic points.[8] One was that the politicians on both sides of the House and the Senate need to work with the President and come up with a sustainable plan for energy. In other words, we need a long-range plan for how we're going to manage energy in this country and where we should be moving. Right now there is no plan like that.

The second point he made was that climate change has become almost a bad word, depending on where you sit on the political spectrum. Some people believe in the science, some people don't. Our strategy thus ought to focus on energy security and the economic impact of using alternatives rather than on what alternatives mean for climate change. That makes sense because if you change to alternatives to bring about energy security and economic prosperity, you'll get beneficial climate change effects anyway.

Daniel Weiss: "Climate change" is indeed a bad word, particularly if you happen to live in a low-lying area in a third-world country. Such people are going to see the impacts of global warming sooner rather than later. In fact, although one cannot link any particular weather event with global warming, the extreme weather we had in 2010 is just a taste of what our future's going to be like if we don't start reducing emissions.

James Trainham said that "some people believe in the science and some people don't." The people who do believe in the science are scientists. The National Academy of Sciences just did a review, finding that of the peer-reviewed papers addressing some element of climate change, about 98 percent agree that as carbon dioxide goes up in the atmosphere, global warming occurs. They also agree that global warming is human-caused. There are very few credible, peer-reviewed studies that lean in the other direction.

Unfortunately, we've seen a political flip of about 180 degrees. It's easy now to forget that 3 years ago, Republican nominee for President John McCain acknowledged that climate change was human-caused and had a plan to deal with it. He even used a cap-and-trade system to address it. Today, cap-and-trade has become like a swear word among Republicans, even though President Ronald Reagan was the first administration to use a cap-and-trade system. It was used to reduce the amount of lead in gasoline in a more cost-effective way. It was used again under President George H. W. Bush when Congress passed the Clean Air Act of 1990. That put a price on the sulfur pollution that was mentioned earlier.

Nonetheless, global warming and cap-and-trade are now swear words, generally speaking, at the national level for the Republican party. In fact, there was an article in a recent *New York Times* Sunday magazine about how the Republican party used to complain about junk science being used to justify environmental laws. Now, the writer claims, the Republican party is, for all intents and purposes, a climate-science-denying party.[9] That's an extraordinary development. Three years ago, global warming was something that the leading candidates, the nominees of both parties, embraced solutions for. Certainly the politics in favor of action on global warming has deteriorated.

During the debate over the previous 2 years on whether to enact global warming legislation, there was a lot of talk about the security elements of the need to act. In fact, the Center for Naval Analysis, which is a nonprofit organization primarily funded by the Department of Defense (DoD), has done a number of studies, led by retired high-ranking military officers, finding that global warming is a "security threat multiplier" and that in volatile parts of the world, it can only exacerbate already existing tensions.

For example, there's greater potential for water wars contributing to the instability that already exists in the Middle East as the impacts of global warming take effect. Certainly, there's been a lot of talk and discussion about it. During the debate in the last 2 years, a group of Iraqi and Afghanistan veterans were very active in trying to create support for action on global warming in part to help reduce oil use in order to increase our national security and reduce the number of U.S. service personnel in harm's way in the Middle East.

Now congressional action on global warming is off the table for the next 2 years, absent some extraordinary event like Miami flooding. We don't need to worry about whether we want to talk about action on global warming with a security message, because there won't be a debate about global warming. In fact, the only present debate about global warming is whether to take away or delay the Environmental Protection Agency's (EPA) existing authority to set emissions reductions under the Clean Air Act that was confirmed under the *Massachusetts v. EPA* Supreme Court decision of 4 years ago.

The debate is not going to be whether to move forward but whether we're going to move backwards and remove authority from the EPA. The debate about investments in clean energy is going to be around economic competitiveness and innovation. That is the framework established by the President for his energy budget when he discussed it both in the State of the Union and in the budget.

Despite his 13 percent cut in the EPA's budget, for example, he would increase investments in energy efficiency, renewable energy, and other clean technology by about a third. This is pretty amazing, given how much he's slashing other areas of his budget. Again, that is going to be a real contrast with the budget that the Republicans just passed through the House. It would cut spending on such things as clean technology, finding ways to wean us off oil, and science, by about $8 billion. The Republicans would cut money from the program to build a new satellite to collect weather data.

That's where the debate is going, and to the extent we continue to have a debate about clean technology, it will be around the need to innovate and to foster economic competition. It's important to note that China is investing $12 billion a month in their clean tech sector. Compare that with what President Obama wants to invest, which is $8 billion for the entire year.

Michael Roberts: I'd like to comment on a couple of matters going back to biofuels and climate change. First, looking at biofuels, it's important not to ignore its implications for food production. When you talk about doing a lot of biofuel production—by which I mean about 10 percent of our fuel, which is 10 percent of our gasoline today at the blend wall— we're using about a third of the U.S. corn crop right now. If the blend wall is raised to 15 percent, you're talking about diverting over half of the U.S. corn crop to fuel production.

The United States is by far the world's biggest producer and exporter of basic food commodities. What happens in the United States drives world food prices. We are by far the most influential country in the world in agriculture. People don't realize that because we take our own food for granted. For us, the price of food—especially the raw commodities—is relatively trivial, so not everyone recognizes the implications. But the ramp-up since 2005 in ethanol production, from almost nothing to about a third of the corn crop, has driven up world food prices.

Since there are many other factors behind price rises, there's a good deal of debate about just how large of an effect ethanol production plays here. It's worth noting, though, that since the production of ethanol has risen very quickly, it's also had the effect of drawing down food inventories. In our own work, we've estimated that it's driven up food prices for the basic staples (corn, soybeans, wheat, and rice) by about 30 percent. In the short run, by drawing down inventories, it makes the world markets a lot more susceptible to other shocks. That's going to be a problem until inventories can be built up to the point that they can meet the new high demand for this kind of commodity. So it's not really clear precisely how much of a role ethanol production has played in price rises, but it's definitely a big factor.

So if one is worried about food prices in poorer parts of the world, say, in the Middle East, this is something to consider. Rising wheat prices have been linked to the uprising in Tunisia and, to some extent, in Egypt as well. The connection between ethanol and food shortage-induced popular discontent is a reasonable inference. There's certainly an element of truth to claims that biofuels production and violence are linked.

Another important thing to think about when considering using a lot of land for biofuels production is that this can cause carbon dioxide (CO_2) expansion not only in the United States, but in other parts of the world. Again, this is something that's highly uncertain, but it's quite possible that if you're clearing more forests or using more land for crop production, you could actually be increasing CO_2 emissions. About 20 percent of the CO_2 emissions in the world are caused, not by the use of fossil fuels, but by the clearing of land, the cutting down of trees, and similar land use changes.

Ethanol production could thus actually be worse than using gasoline when it comes to causing CO_2 emissions. In other words, there are externalities that ought to be considered here, too. In our ideal Pigouvian world, we'd want to put a price on carbon, both sequestrations and emissions. You would want to do both. This is fine in theory though hard to implement, but I still support it.

Of course, sequestrations can be a little harder to price, and that can get fairly controversial. The strategies for dealing with that so far are very challenging. The idea is to pay people for offsets.[10] There's a bit of a market for carbon offsets right now, but it's very difficult to make it work effectively. However, there are lots of economists trying to understand them better.

It's also important to note the effect of climate change—the potential effect of climate change—on production. I'm among the pessimists. The statistical evidence is very powerful, showing that climate change, if it happens as projected, could have an absolutely devastating effect on crop yields in the United States and probably in other parts of the world. We've got the best land and climate in the world here in the United States. If it's going to get a lot worse here, it's certainly going to get worse in other places, too. Globally, the change in climate doesn't bode well for crop production.

To what extent are there offsetting factors from the changes in temperatures and precipitation? Are there potential positive effects from CO2 fertilization?[11] Years ago, most of the scientists thought that this was going to be a boon, that the CO2 fertilization effect was going to be greatly beneficial, outweighing any effects from temperatures and thus a boon for food production. But that's changed. These days, the scientists are much more skeptical about the benefits of CO2 fertilization, and the negative effects from extreme heat are looking much worse. Yield growth and productivity growth in agriculture have significantly slowed in recent years.

There is cause for concern about what's going to happen to food prices. Again, not for us; we're not going to notice it in the price that we pay for food at the stores, but for basic staples and for the poorest two billion people in the world, people living on $2 a day or less, they'll really feel it. They did feel it in 2008, and they're starting to feel it right now. It can very well become a security problem.

Daniel Weiss: In fact, food prices hit a world record yesterday.

James Bartis : I want to endorse the comment that only nonscientists disbelieve that climate change is a major problem. Basically, every major U.S. scientific organization with expertise in this area has come out very strongly supporting the view that it is a problem. The National Academy of Sciences and the American Association for the Advancement of Science came up with a very strong finding that man-made emissions are driving climate change more rapidly than would otherwise be the case.

There are a few scientists who have raised some doubts, but they are a minority. They might be right. But even if 50 percent of our scientists are wrong, which is a ridiculously high figure, you would still not sit and do nothing. So it's important from a global security point of view that we do something. There's so much low-hanging fruit that can be addressed at negligible costs, and there are health benefits and other environmental benefits associated with taking some action.

I mentioned the fact that the gasoline tax is inadequate to pay for the highway infrastructure. Raising the gas tax to pay for the roads and bridges would also result in lower greenhouse gas emissions.

With regard to energy, using the security argument to promote reduced use of petroleum doesn't get us very far. There's a tremendous amount of fossil fuel out there. There's enough petroleum and fuels, in my view, to toast the atmosphere three times over. It will cost a little bit more, but there's no shortage of fuel. So the real driver, to me, is climate change. That and the general environment: the particulate matter and the soot that are going into the air. Those are the big arguments for our doing something about energy and energy security.

If we didn't have climate change problems, we could solve a lot of our problems just by relying on coal, for example. We can use coal to make liquid fuels. We've got huge amounts of oil shale in our country, three times the size of the equivalent reserves of Saudi Arabia. We could use those. They don't cost much more. I think they're probably economic at today's prices. So we don't have a problem if we're not worried about climate change.

David Dayton: I would say that as long as oil stays upwards of $100 a barrel or higher, then a lot of those alternatives like coal and tar sands look pretty good. But if the oil supply starts to increase, the spigots open a little bit more, the price goes back down to $40, $50 a barrel again, then we're going to have the same cyclical argument that we've always had. It's not as bad as the 1980s, when the radical drop in oil prices killed all the alternative fuel research. But the economics of the alternatives are still going to be higher than petroleum-derived fuels.

There's got to be a way to underwrite the development of these alternative technologies so that consumers can choose between petroleum and something else. Right now we don't have a choice. For transportation, it's uni-fuel. We've got one option and one option only. For electricity, we've got several options, but the economic benefits of one clearly outweigh the others. We talked about nuclear energy in the last panel. If other externality factors are not factored in, nuclear power has $4,000 or $5,000 a kilowatt hour installed capacity while natural gas has $350 a kilowatt hour of installed capacity. Given that, the decision to use natural gas is a no-brainer. I think that's the argument for petroleum right now. If petroleum's $3 a gallon, the alternatives are not viable. There's got to be some sort of mechanism for getting alternative competition into the market.

James Bartis: If we factor the externalities into the cost of fuel, we wouldn't have what we have today, which is Congress saying this chemical or this particular fuel gets a subsidy, this one does not. That way of doing things is based more on transferring wealth to certain parts of the country and can happen just because there are two senators who push it in one state, or there is a low population in another. If, instead, you set up a framework so that the marketplace can make the decisions, I think we're going to be far better off. I'm optimistic that as we try to address our deficit and reduce government spending, we might consider taxing fuel use. We could do that instead of continuing to tax incomes. It would not necessarily have a negative net impact on GDP or development.

David Dayton: I don't think that's going to get you elected.

James Bartis: We're not going to get out of this budget problem that we have without raising some revenues.

Michael Weiss: In 1993, the House of Representatives did pass a BTU tax as part of President Clinton's economic recovery package. It hit some rocky shoals in the Senate, and they replaced it with a 5-cent-a-gallon increase in the gasoline tax, which is where it's remained since then. It's been 18 cents a gallon since 1993. In this political environment, I think anything with the word "tax" in it is going to be something that a large number of members of Congress are going to avoid, regardless of whether or not it makes economic sense.

Daniel Roberts: I completely agree with the idea of gas taxes, but I don't believe it's going to happen anytime soon.

James Trainham: It's one thing just to add the tax. It's another to trade it off. Can you do that? That's the question. Right now, as I pointed out, one of the externalities is the huge military cost we have to keep oil flowing out through the Gulf of Hormuz, and we do that for the whole world. Nobody else pays that tax except the American taxpayer. If we had alternatives, we could pull the military back. But we don't. And you'd have to replace a significant amount of oil use to change the U.S. energy policy.

The question is really whether we can convince people on other grounds of the need to develop alternatives, and that means looking at some of the things we talked about like healthcare costs and the cost of infrastructure. If we change fuel taxes to reflect the real cost of fuel but take the income tax down so the net tax is essentially the same, some folks will complain. They will argue, "Oh, the poor people are being harmed—this is a consumption tax." But that can be taken care of.

Michael Roberts: I don't think the obstacle is necessarily the people. You actually could inform the public of the advantages of doing this. I think they might buy into something like this, or at least enough of them would, especially if you talked about reducing income taxes. But you're facing some real political headwinds because there are some powerful oil interests that are going to be fighting against you. It'll be the monied oil lobby and allied interests in favor of oil that you'll be fighting against.

Daniel Weiss: Two weeks ago, the House of Representatives voted to block EPA from setting reduction standards for mercury and other toxic chemicals coming from industrial boilers. That's what we're talking about here. Outside of some extraordinary event, it's extremely unlikely that they are going to do anything to raise revenue through any sort of petroleum tax at least over the next few years. Politically, the idea is a complete nonstarter.

It's true that the American people by overwhelming margins support EPA regulating global warming pollutants. They do so if even if they're asked in the worst way possible. Opinion polls show that Americans still favor it by a 20-point margin, even if it costs jobs or raises prices. The problem is in American politics where a vocal minority will beat a complacent majority almost every time, and the vocal minority right now is on the side of do-nothing.

Even though there already is public support, you have politicians who care little about health impacts and who, in the name of jobs, are going to try to undo or block other clean air health standards to appease their constituents or to please their funders. Take, for example, the Koch brothers and the American Petroleum Institute (API). They are now saying they're going to be directly contributing to campaigns, something they've never done before.[12] So I think the odds are pretty low that anything positive like a fuel tax will happen.

James Bartis: ExxonMobil came out in favor of a carbon tax.

Daniel Weiss: Yes, but did they spend any of their money lobbying in favor of it? No. In fact, they give money to the American Petroleum Institute, which helped run $100 million worth of ads attacking the existing global warming legislation. I think that Exxon's position on the carbon tax is more fig leaf than policy.

SHALE GAS: A KEY TO ENERGY SECURITY?

James Trainham, James Bartis, David Dayton, Michael Roberts, and Daniel Weiss

I'm now going to address another subject—shale gas. This isn't strictly one of the renewable energies, but I think it's such an important part of our quest for energy security that we should not omit it from the discussion. It's likely to play a role as we transition to alternatives. How might we factor shale gas into our energy future? Shale gas is going to become quite prevalent, and we have huge reserves. The latest assessments show that reserves of natural gas went up by about 39 percent, and it will go up that much again by the time of our next assessment in the summer of 2012.

Natural gas can provide us with an interim solution before we can develop some more of the true alternatives and build the infrastructure we need to make those a reality. I think we all agree that we have alternatives to oil when it comes to electrical power, among them shale gases. Can we replace oil with the natural gas and gas-to-liquids technology? We've done this in the petroleum industry and the chemical industry for a long time. It's not an unknown technology, nor something which calls for a great invention. With more experience, we could bring the cost down. So I'd like some comments on a potential move to use shale gas as a source for gas-to-liquids transportation fuels.

David Dayton: One of the challenges of that is, again, the existing infrastructure. That's because 80 percent of the world's methanol today is made from natural gas.[13] The current figure might be a little higher than that. As we heard earlier today, methanol is not a functional fuel at this moment, but we can convert our cars to burn it as a fuel. Top Fuel dragsters and Formula One racers use methanol for fuel all the time. It could be a fine fuel for our autos. So I think one of the challenges with gas-to-liquids is trying to make something that's hydrocarbon compatible.

You can use methanol-to-gasoline (MTG) technology.[14] It's almost but not quite gasoline. It's highly aromatic, so it's a blend stock. It's not going to replace all of your petroleum. Although the Fischer-Tropsch plant in Qatar is having technical troubles right now, the process is technically feasible.[15] It's just not terribly cost-effective. Another option for shale gas is to use it in compressed natural gas (CNG) vehicles.[16] You don't have to do anything to it except compress it and put it on a vehicle. All the buses in Chapel Hill run on CNG right now or as hybrids.

James Bartis: My information is that Qatar Petroleum's plants are not in trouble technically at this time. Right now, the world makes about 300,000 barrels a day of petroleum substitutes using a process called the Fischer-Tropsch process, which was invented in Germany in the 1920s. The plants the Germans built were frequently bombing targets of the United States and the Allies during World War II. When South Africa went apartheid in 1954, it was worried about being embargoed, so it built a Fischer-Tropsch plant that used coal. More recently, a second generation has come about. Mobil built a plant in New Zealand that converts natural gas to high quality gasoline.

They shut the New Zealand plant in the 1990s because the process, being energy-intensive, was not competitive with the low oil prices that were prevalent during that period. More recently, Shell built a plant in Malaysia that's making 17,000 barrels a day. A couple of years ago, a brand new plant started in Qatar making 35,000 barrels a day. Shell is starting up another plant in Qatar that is designed to make 140,000 barrels. So by the end of this year, we're going to have almost 200,000 barrels a day of production from this method based on natural gas. The method works only if the natural gas input is fairly low cost, and the one place where there's going to be low-cost gas in the United States is Alaska.

Audience: Prudhoe Bay?

James Bartis: With all the shale gas coming in, I think the likelihood is small that we'll build a pipeline from Alaska through Canada to the United States to bring Alaskan gas to a low-cost gas market. There is a divergence between what natural gas costs and what oil costs. Wholesale gas today is going for the equivalent of $24-a-barrel oil.[17] If you can take that gas and make it into an oil product, you can sell it at $80 or $90 a barrel today. That's the motivation. You are most likely to see that happen in Alaska. Whether that makes sense elsewhere in the United States really depends on whether there are other opportunities to use that natural gas. We've lost a lot of our chemical industry to Saudi Arabia because of the high price of gas here. These industries might revive due to the availability of gas from shale formations in the United States.

Daniel Weiss: Fuel produced from coal has a much higher carbon dioxide content than regular gasoline. Is that true for fuel produced from natural gas as well?

James Bartis: Actually, the coal-based fuel has a lower carbon dioxide content than regular gasoline.

Daniel Weiss: Are we talking about gas or coal?

James Bartis: Either one. But what really counts is life cycle emissions. If you start with coal and examine life cycle emissions, you double greenhouse gas emissions compared to conventional oil. So if you're worrying about global warming, that's an unacceptable approach. There are ways to use liquid fuels from natural gas without carbon sequestration in such a way as to break even with conventional petroleum. By that, I mean that you will end up with a fuel that has 5 percent higher or 5 percent lower carbon dioxide content. That won't change the world. There's a lot of hydrogen in natural gas. There's not in coal. That's why you end up with very high emissions when you use coal. So this is not a greenhouse gas problem. It doesn't fix the problem, but it doesn't make it any worse.

David Dayton: Shale gas won't be cheap. There aren't many refineries being built in this country, so even from a peak oil perspective, we may be refinery limited, not production limited. But the point is that if you put a natural gas-to-liquids plant in, you're going to have to invest upwards of $3 to $4 billion, depending on the size of the plant, to make it economically competitive.

Daniel Weiss: The Center for American Progress has been a long-time advocate of increased use of natural gas as a bridge fuel to a clean energy future that's probably 20 or 30 or more years away. A recent *New York Times* report found that the production of shale gas has a lot of potential and actual environmental impacts. The series did not even address the fugitive methane that escapes when doing the hydraulic fracturing that produces the shale gas. Those issues are going to have to be addressed concurrently because otherwise we will address one problem while creating another. So far, the natural gas producers have resisted any sort of federal oversight and standards for groundwater and surface water protection, air pollution, and methane capture.

James Bartis: Shale gas has happened so quickly that the states and federal government have been unable to move in terms of regulating it.

Daniel Weiss: Right now, thanks to the 2005 energy law, the federal government is prohibited from regulating shale gas's impact on drinking water, but it still has authority over clean air and clean water.

James Bartis: To mitigate these problems, the price is going to have to go up. But I don't see that stopping shale development. It just means that instead of coming in at $3.50, the cost might be $4 or $4.50. That's the price we pay to live in a place where we can drink our water, unlike China, for example.

ALTERNATIVE ENERGY AND INTERNATIONAL RELATIONS

James Trainham, James Bartis, David Dayton, Michael Roberts, and Daniel Weiss

Let's turn to the future. What are the implications of a new world where alternatives might begin to have a significant impact on our energy picture? How would this affect our relations/political interactions with OPEC and other oil-producing countries?

James Bartis: The International Energy Agency (IEA) completed a world oil study. They took a look at the world oil supply and demand in 2035, modeling one scenario in which governments took action to reduce greenhouse gases. As a result, petroleum demand was on the order of about 90 million barrels per day, and the price of oil was fairly low. The alternative scenario was one in which governments didn't do anything about global warming. Then the demand was for about 125 or so million barrels a day, and the price was about $150 a barrel.

It's possible that in the future we will reduce dependence on OPEC oil, but the only way this will happen is if we and other countries put some additional cost on the oil we use. According to one scenario, we'll just continue the way we're going. In that case, the production of oil globally will just keep going up, a greater and greater share will go to OPEC, and prices will get higher. According to another scenario, we'll do something dramatic about global warming. Countries that are oil users will place a tax on oil and reap the social benefits. We'll use alternatives, and it's not just alternative fuels. There is also a large role for energy efficiency and major changes to our infrastructure, changes to how we transport goods and how we move around, and where we live.

When we think of dependence on Middle East oil, we should think what this term really means. Asia gets 75 percent of the oil that goes through the Straits of Hormuz. Only about 12 percent comes to the United States. Often people will say "Well, let's get the United States independent of the Middle East." Actually, that could be easily done. We could pass a law tomorrow saying that in future we wouldn't import any oil that comes from these five nations in the Gulf. We could get our oil from somewhere else. It might cost a little bit more because of logistics, but not much more. But would that change anything? We are in the Middle East because of Iraq, because of Afghanistan. How would that change things? Our allies and China are still going to depend on that oil, even if we are not. I don't think that there is a big security externality here. If there is, it is because we don't want other people to play over there.

James Trainham: When it comes to our military occupation, my question is: Do we have a big military presence in Tahiti, and why don't we? They don't have any oil. Why would we want a military presence in the Middle East if they didn't have oil?

James Bartis: We would have a large military presence there regardless. I don't think our presence there depends on the fact that we are importing oil from the region. It is a strategically important place for the entire world.

David Dayton: But do we have a military presence in Venezuela?

James Bartis: Hopefully we will continue not to.

David Dayton: We still get oil from Venezuela, too, so the two are not mutually connected.

James Bartis: The Middle East is a little bit more unstable than South America. But it doesn't mean that we have to be the only presence in the Middle East.

Michael Roberts: Extracting oil in the Middle East is very cheap, and while supplies are being depleted, a lot remains. If the United States and/or other countries move to alternative energies very quickly and if the demand for oil is reduced, the price of oil will go down. So you're going to have these constant competitive pressures. In other words, you can't just look at the price of oil today and say, "Oh, once oil prices reach $100 a barrel, we can come up with alternatives." If you come up with those alternatives, the price of oil will drop and compete with your alternatives, and it will compete with your alternatives all the way down the line. The Middle East will still produce quite a lot of oil. So to really turn off the taps, you do need to come up with a true miracle, something that is far better than oil in many different ways.

That leads me to one last point. There are positive as well as negative externalities in the world. For example, coming up with new ideas and new technologies is important. Yet there's always underinvestment by private markets in research and development because it's hard to protect your ideas perfectly. So one thing you can do—and it's perhaps more politically feasible than taxing negative externalities—would be to subsidize these kinds of positive externalities.

Daniel Weiss: It seems to me that you need three things for developing any sort of clean technology, whether it's for alternative fuels or electricity. You need to have markets, finance, and infrastructure. The problem with just investing in research in technology is that you also need to build a market, and you need the infrastructure to deliver the new product.

The Brookings Institute produced a report in 2011 that basically said, "Let's just put 20 billion or so into research, into research and development on clean energy technologies." That would be valuable, but without helping create a market for those technologies, and without having the infrastructure to deliver them, it will be very difficult to convert those innovations into commercial scale technologies.

James Bartis: That's why it's so important to assure a price differential that reflects the costs of the negative externalities. That's crucial. If you want to stabilize innovation and market penetration, you've got to have a stable price differential structure in place.[18] That calls for putting a tax, or whatever you want to call it, on fossil fuels, whether on carbon or infrastructure or a combination of both. We also can't dismiss the importance of energy conservation. A barrel saved is a barrel produced. Energy conservation and energy efficiency are powerful tools.

Daniel Weiss: About 30 states now have either renewable electricity standards or "a renewable portfolio standard." Utilities in these states are required to produce a certain amount of their electricity from renewable sources. There is such a standard in North Carolina. So, too, in Pennsylvania, but there you can also use coal ash or something else. The idea is to help create a market for new clean energy technologies without spending state resources or state taxpayer money. California was one of the earliest states to do this, and it adopted one of the most aggressive renewable electricity standards. Now there is a homegrown clean tech industry in California. The standards helped create the market. One of the things that the President proposed in his State of the Union address is what he calls a "clean energy standard." This would include not only true renewables but also nuclear power, natural gas, and electricity from coal plants that employ "carbon capture-and-storage" technology.

But that works only if you have two tiers—one for true renewable electricity and one for low carbon electricity. We don't want cheap natural gas (at least early on) competing against wind and solar. We need a target for true renewables as well as a separate target for so-called low carbon energy that will help create a market. We need to provide funding for the research, development, and initial deployment of renewable technologies. We need to develop and enhance the infrastructure to deliver the electricity. That's the positive vision. I've been sitting through these panels, and each of them has been more depressing than the one before it. There is a positive vision, and it's being written in many states right now. Unfortunately, it's not going to be written at the federal level in any sort of comprehensive way for at least the next 2 years.

James Bartis: It may be a positive vision, but every analysis I am aware of suggests that it's one of the least economically efficient approaches you can take. It is a secret tax, and that's why it's so popular among our politicians. When they set standards, they can force you as rate payers to take a particular course, but the politicians never have to say, "Here's what it's going to cost you." In fact, almost none of these decisions at the state level are based on cost calculations. When government establishes tiers and picks technologies, citizens and businesses are often forced to make decisions that are very inefficient. If the goal is to manage CO_2 emissions, then put a price on CO_2 emissions.

Daniel Weiss: We already pick winners and losers. We've had oil subsidies for 100 years. We had coal subsidies before that. We've been investing in nuclear power since 1948. So we've always picked winners and losers. Now what we need to do is think of other values in addition to economic efficiency. A renewable electricity standard helps create a market for these emerging technologies. We need the finance, we need the transmission. Eventually, those technologies will get to an efficient scale. Right now, for example, the cost of wind energy—and this was not the case 10 years ago—is only 4 to 5 cents per kilowatt hour.

Audience: When the wind blows.

Daniel Weiss: It's an intermittent source of power, that's absolutely right, but the point is that the price has come down dramatically. That's because renewable electricity standards have helped create a market for these technologies, and establish economies of scale. If it hadn't been for these standards, these new clean technologies wouldn't have gotten off the ground after we've subsidized their competitors for 60 to 100 years.

James Bartis: Wind is still not economic in almost all parts of the country. It blows primarily at night. That's not when you want it.

James Trainham: It's mostly in the Midwest.

Daniel Weiss: We've got great advantages in this country. The Department of Energy (DoE) under Bush said we could be at 20 percent wind-generated electricity within 20 years because of our vast wind resources, not only onshore but offshore.

James Bartis: The problem is, again, it's the government picking technologies. We would be much better off if we got the government out of this calculus and simply set up a broad-based economic and policy framework so that we as citizens are motivated to use the lowest cost fuels that have the kind of benefits that the government seeks.

Daniel Weiss: I think that's a great idea, and as soon as the oil companies and the nuclear companies give back all the money they have received from the federal government over the past 100 years, we will have a level playing field. But until then, we have an economic interest and security interest as well as an environmental and public health interest in generating new cleaner technologies. Therefore, the government can play a very important role in not only providing the seed money or financing for research, development, and deployment (RD&D), but also in helping to create a market. It's not a level playing field.

QUESTIONS AND ANSWERS

Q: We all know if we had a choice, we'd rather just produce hydrogen because it's not going to increase our carbon footprint, and water is the only by-product of burning hydrogen. Water is a pretty easy by-product to live with. Anything else that we burn usually puts carbon dioxide in the atmosphere. It is not a good thing to burn food crops and it is not a good thing to produce food crops to burn for food. But surely it is a good thing to use products that cannot be grown on land that can be used for growing food? How long does it take for carbon dioxide in the atmosphere to go down into the earth and produce oil or natural gas? We are talking about taking a plant that's growing on property that doesn't support food and combusting it or converting into some sort of a fuel. How long does it take that to reabsorb into the plant?

Michael Roberts: I understand that, but I think when you're talking about using a lot of land, you're going to be competing against other uses of that land.

Audience: No, there's a lot of land in the country that doesn't support food cropping.

Michael Roberts: It supports something. There are a lot of challenges to overcome just in terms of transporting the kind of volume that you need. If you overcome those obstacles, then you have to look at a particular example more carefully to see what the feedback effects would be. It's possible. Ethanol came on line when corn was $2 a bushel or less, and people thought it was great. It was great at $2 a bushel. But people failed to recognize that these prices change when you increase demand that much.

Audience: I want to question Mr. Roberts's comments about ethanol and the impact it has on the corn crop. It's my understanding that basically all of the corn grown in the United States is grown for cattle for meat consumption, and it's the increase in demand for meat around the world in the growing economies that's affecting the price of corn. Plus global climate change has affected the weather and uncertainty of crops. Then there are the petroleum inputs that are going into our agriculture and the corn crop. All these things are affecting the price of food rather than ethanol itself.

Michael Roberts: Ethanol's one big factor. One-third of the corn crop is going into ethanol right now. That's a pretty big number.

Audience: Does that take into account the fact that some dry distiller grains are actually going to feed the animals?

Michael Roberts: That offsets it by about a third, between 20 and 30 percent, so some production goes back to feed, but not much.

Audience: I'm really trying to straighten this out because I do promote the use of alternative fuels and corn-based ethanol in particular, and that's been something I've found a bit troubling.

Michael Roberts: We're probably getting agricultural land expansion throughout the world. We're estimating that in lots of places this is the result of growing crops for ethanol. But it's hard to tell where the land is really coming from. We don't know if it's really causing deforestation, though I wouldn't be surprised if it were.

Audience: However, it's my understanding that a lot of the land expansion is cleared for cattle, and it's really meat production that's driving this, not the corn. We're not eating this corn. So really we should be talking more about cutting our meat consumption rather than getting rid of corn for ethanol.

Michael Roberts: Cutting meat consumption would help a lot.

Audience: But do we really want meat police out there?

Audience: I'm not suggesting all of us become vegetarians.

Michael Roberts: I don't disagree with you. All of these are relevant factors. Dietary demands are a factor. China's a big part of this, too.

David Dayton: We're talking about the expansion of the corn ethanol industry. To look at increased volumes of the industry right now, we're at 12 billion gallons per year, and the food, feed, and fuel are sort of in balance. At least there's enough food grown to be exported and enough grain to feed the cows and enough ethanol to go into the fuel market.

Michael Roberts: The corn farmers would love to have prices even above $8 a bushel. Prices are linked to demand. But the thing about commodity crops, and all the staples pretty much, is that supply and demand are both highly inelastic. That means that very small shifts in supply or demand can have a huge effect on price. But we in the United States won't notice that.

Prices could go up a lot. We could push prices for corn up to $30 a bushel, and maybe the price of a Big Mac would go up a dollar. Do you think people are going to eat much less meat? Probably not, but a lot of people would be starving in the world because when corn prices go up, soybean prices go up; when soybean prices go up, wheat prices go up. If you look historically, those prices move almost in tandem and usually rice isn't too far behind. All of these prices are very connected. Our estimates are that the growth in ethanol since 2005 has caused world commodity prices for these staples to be about 20 to 30 percent higher than they would've been without it.

Prices these days are about three times what they were then, so of course there are lots of other factors at play, but that's not our sole interest. Because ethanol was brought on-line very quickly, it drew down inventories. As inventories get low, then markets become extremely susceptible to other shocks. They are more sensitive to surprises in the weather. Until we get more experience with this new market in ethanol, we will be susceptible to such shocks.

James Bartis: The other big issue here is that the demand for food is rising, and every time you clear a carbon-rich ecosystem of vegetation so that you can produce food, a large pulse of greenhouse gas emissions enters the atmosphere. Scientists have examined that, and it really depends on where you are with respect to the margin, but in some cases, e.g., a rainforest, it takes hundreds of years before you break even on the emissions.

We're worried about greenhouse gas emissions over the next 40 years. Taking land out of its current use, whether it's in the conservation reserve or it's prairie land or whatever, and directing it toward energy, is not going to be productive in terms of greenhouse gases. I think there's a general consensus that we're not going to expand corn ethanol beyond the current standard of 15 percent that Congress has set.

David Dayton: Renewable Fuel Standard 2 (RFS2) caps corn ethanol at 15 billion gallons a year.[19]

James Bartis: We're already at 12 percent, so we're almost at the limit here.

Daniel Weiss: RFS2 also requires 22 billion gallons of advanced biofuels by 2022. Do you think that target will be met or do you think that is completely unrealistic? Or do you think it's somewhere in between?

David Dayton: There are two ways to look at that. One, EPA has been downgrading the current cellulosic ethanol requirements for RFS2 since it was enacted because plants haven't come on line and been productive like they were supposed to when the law was passed.

Daniel Weiss: You mean to meet interim standards?

David Dayton: Yes, to meet interim standards. There are yearly targets, and so far, I don't believe we've hit one. If you look at the capital expenditures for 22 billion gallons per year, current advanced biofuel technologies are going to be on the order of—assuming you get a really big optimized plant—200 million gallons a year. So by 2022, you'll need roughly 350 plants to meet the 21-billion-per-year requirement?

Right now, according to the literature, current biofuels plant design costs are on the order of $2 to $3 hundred million per plant. Let's assume the early models are on the order of two to three times that, so we have a billion dollars a plant times 350—that's a lot of capital that needs to go in the ground, a lot of steel, mortar, concrete that needs to be put in, before that even happens.

If you look at 2022 and start working backwards, first of all, the technologies aren't quite ready for prime time yet, and we're already in 2012. So from 2022, let's say it takes 2 years for the permit process, give or take, so now we're down to 2020, and 3 to 4 years to build the plant, so now we're down to 2016, so that means in 4 years, we've got to have 350 plants in the ground ready for commissioning.

James Bartis: And no experience base in existence.

David Dayton: We haven't figured anything out yet.

James Bartis: These are all first-of-a-kind plants.

David Dayton: So that's my response.

James Trainham: So your response is, "It's not going to be possible."

Daniel Weiss: However, the good news is that because of this mandate, we're moving faster in that direction than we would have otherwise, correct?

David Dayton: Yes, at least we're still talking about it because there is a mandate, because there's climate change, because there are energy efficiency and energy security issues, because alternative fuels haven't been pushed aside, and now oil's $100 a barrel. We have all these reasons to at least keep talking about it now.

Comment: My name's Ward Lenz. I'm Director of the State Energy Office here in North Carolina. I want to give two responses in regard to the Renewable Portfolio Standard (RPS) the renewable fuel volume mandate. First, in North Carolina, we do have a portfolio standard for renewable energy and efficiency. But, contrary to the gentleman's point, we actually do have a price cap. So if the price of those renewables gets too high for the utilities, then they are capped at a certain level.

Second, my friends in Iowa would be angry with me if I didn't talk about ethanol. The fact is that when you make corn ethanol, only the starch is taken out of the corn, and that's what you make the ethanol out of. There's actually very little of the nutrient value of that corn that is lost in the process because you get dry grain or corn gluten meal, which then generally goes to the pigs and cows. In other words, 75 to 90 percent of that food product is still in the market. The food-versus-fuel question thus looks a little different—the amount of food lost is a lot less when you look at it this way than when you say a third to half of the corn product goes into an ethanol plant.

Q: One thing the panel has not talked about much is the impact we could make on energy through conservation. We have 100 million homes in this country that all need to be weatherized. These lights in this room, these fixtures could be replaced with LEDs. I'm told we could reduce electricity use by 20 percent by doing nothing but replacing lighting in this country. Could you comment on the impact of conservation?

Daniel Weiss: That's a great point. In 2007, as part of the Energy Independence and Security Act, Congressman Fred Upton of Michigan put in a provision that would increase the efficiency requirements for light bulbs that would basically lead to a phase-out of incandescent light bulbs. Last year, when he was running to become the Republican chair of that committee, Rush Limbaugh attacked him for that because it suggested a "nanny" state. In other words, we have a God-given American right to waste energy if we want to.

Mr. Upton then said, "Well, we're going to have hearings on this now." In other words, he was walking back from his own provision. One of the things that the President's budget would do is to increase investments in energy efficiency. There's already a big low income home weatherization program that got a very slow start under the American Recovery and Reinvestment Act. It will weatherize five million homes. In addition, we've been working with both the White House and Republicans and Democrats in Congress to pass what's called the Home Star program that would create an incentive for homeowners to retrofit their own homes. The weatherization program goes to low income homeowners. Home Star would be for middle class homeowners. It would also help create jobs at places like Home Depot and efficiency material manufacturers. It passed the House in 2010, and even some very conservative members of the House supported it last year.

It's not clear right now whether Congress is willing to spend the $6 billion, which would generate about $50 billion of private investment. We're very much in a budget-cutting mode regardless of whether we're talking about cutting wasteful spending such as building a Lawrence Welk museum in North Dakota or talking about investing in future technologies. I totally agree with you that energy efficiency has got to play a key role. States like North Carolina and others that have an "energy efficiency resource standard" are blazing the trail.

Audience: If you want to drive Americans to conserve, give them a financial incentive. I put in a tankless water heater in December and got a tax credit.

James Bartis: One of the key tools in all this is to set efficiency standards a lot more aggressively than we have in the past. There are other countries that have good systems in place, and we can copy their successes. Look at Japan.

Audience: And California.

James Bartis: Japan has a very good system. They have building standards and building codes for remodeling. There are many opportunities to do this that we're not taking advantage of.

Daniel Weiss: California was mentioned. There is one possibility for some progress at the federal level, and it is in the area of building efficiency standards. This has bipartisan support, partly because it doesn't cost the government anything, and because it's an inducement for the states to upgrade their building standards. This measure could conceivably pass the Senate. My guess is it will still have a tough time in the House because people like Rush Limbaugh, Glenn Beck, and those of similar opinions will again attack it as being part of the nanny state. But it is a possibility down the road for bipartisan cooperation.

James Trainham: Let us comment on the architects who have a 2030 goal. Reliable estimates are that 60 percent of our energy is used for heating and cooling buildings. There is also a significant amount of energy required to produce the materials to build new buildings. When you take all that into account, you solve a significant portion of the energy problem if you solve the energy efficiency issue. The architects have set a goal by 2030 to come up with building techniques, as well as retrofitting techniques, that would enable you to have zero or net energy buildings. DoE is pushing this. It's focusing on energy efficiency and building efficiency. There's a real effort going forward in this area, and all of us would likely support it.

James Bartis: I just participated in an energy review at Fort Bliss, Texas. The Army standard is 30 percent tighter for new buildings than the national code. The experts from the Army Corps of Engineers believe the Army's standard needs to be tightened by a factor of two. Furthermore, that's already below the commercial standard. We're talking about an improvement of at least a factor of two in building energy efficiency. That's a huge amount.

Daniel Weiss: The entire Department of Defense is committed to building either silver or gold leadership in environmental energy and design (LEED) certified buildings for all new installations.[20]

James Bartis: Silver. They should be gold, but it's silver right now.

Daniel Weiss: That's a big start.

Q: I'd like to revisit the earlier discussion of ethanol and food. There are studies all over the map on this issue. Let me make two points. First, if you look at the revised World Bank study by Baffes and Haniotis, looking at the 2006 through 2008 commodity price boom, they concluded that what was driving the food price boom was primarily oil price and speculation, not biofuels production.[21] Indeed, if you look at that whole period, you'll note that U.S. food exports increased throughout.

Second, we need to keep in mind the fact that ethanol finally made corn economic. We don't pay corn price supports anymore. Before, we were supporting cheap corn with taxpayer money and effectively dumping that cheap corn on the global market, which was undermining economies in the poor and developing world.

Michael Roberts: That's a classic line that we often hear, but even without the subsidies, we'd raise the same amount of corn. Since 1996, most of the subsidies have been decoupled from whatever farmers have decided to produce.

Michael Roberts: But recently there was a short period when corn prices and oil prices were locked together. This was particularly the period before we hit the blend wall but when we were still ramping up ethanol production very rapidly. The prices got locked together because ethanol was a substitute for gasoline at that point. However, ethanol wouldn't have got started if it hadn't been for the subsidies.

Audience: Yes, because of the subsidies.

Michael Roberts: The only reason they built all of the ethanol plants in the first place was that they had the implied mandate and a 50-cent-per-gallon subsidy to engage in it. But the subsidy was not "economic" in its effect. Enriching farmers even more than they already are was not a necessary inducement.

Audience: Even if you took away the blender credit and even if you took the tariff off imported ethanol, as long as oil is above $55 a barrel, you are going to have the big producers turning out ethanol because it's economic. Moreover, you can use it in a vehicle—even if it's not flex fuel—up to 10 or 15 percent.

Michael Roberts: All that is true now that the ethanol plants are there. But what got them into the game to begin with? Now that the ethanol plants are there, even if you took away the subsidies, I'd agree with you that ethanol's here to stay. It's not going to go anyplace. But it's a very different question and highly uncertain to say that we'd have ethanol at the scale we have today, or anywhere near it, if it wasn't for the subsidies that were put in place back in 2005.

Q: I'll start with two premises, then ask a question. The first premise is that petroleum is our real energy security issue, and the second premise is that carbon dioxide is not the issue that's going to drive any kind of tax or price support program in the near future in this country. The question: Is there a technology-neutral incentive, like an alternative fuels incentive program, that you would hold up as an example? That is, is there a country somewhere in the world that has a program we should emulate? I'm thinking of alternative fuels, not carbon dioxide-driven. What kind of incentive along those lines could be put in place?

Anne Korin: There are two examples. First, obviously you have Brazil. We saw how well that worked in 2008 because when oil prices spiked, a significant portion of the Brazilian vehicle fleet was immune because it used flex fuel. Consumers compared the per-mile price with that of gasoline. You have different energy densities, so you have to look at the per-mile price, not the per-gallon price. Without subsidies, alcohol was actually more economical. That year, in 2008, gasoline became an alternative fuel in Brazil. If you let the market work, it'll work. In another year, depending on the relative prices of the different underlying commodities, you might get a different result.

The other example is China. You are seeing a trend there, which started in Shanxi province, their coal province, that is now moving to three other provinces. This trend is toward a very drastic expansion of coal-to-methanol conversion. If that trend is on target, 50 percent of world methanol is going to be made from coal, primarily driven by Chinese production. They are gasifying that coal and turning the gas into methanol. At the same time, the auto companies are producing flex-fuel cars that can handle methanol. You're seeing China move in that direction to Standard Specification for Fuel Methanol M70-M85 (M85) standards across the different provinces.

ENDNOTES - CHAPTER 4

1. Nicholas Gregory "Greg" Mankiw is an American conservative macroeconomist and Professor of Economics at Harvard University known for his work on New Keynesian economics.

2. A Pigovian tax is a tax levied on a market activity that generates negative externalities. The tax is intended to correct the market outcome.

3. The blend wall is the maximum percentage of ethanol that may legally be added to gasoline.

4. E85 is an abbreviation for an ethanol fuel blend of up to 85 percent denatured ethanol fuel and gasoline or other hydrocarbon (HC) by volume. E85 is commonly used by flex-fuel vehicles in the United States, Canada, and Europe.

5. Advanced hydrocarbon fuels are derived from cellulosic biomass and can be used as direct replacements for gasoline, diesel, and jet fuel.

6. Camelina is a genus within the flowering plant family Brassicaceae. One species, Camelina sativa, is a historic and potentially important oil plant. Jatropha is a genus of approximately 175 succulent plants, shrubs, and trees commonly known as physic nut. Jatropha contains compounds that are highly toxic.

7. It may be possible to grow algae using land and water unsuitable for food production. Algae consume carbon dioxide (CO_2) as they grow, and produce bio-oils through photosynthesis. This bio-oil can be hydrotreated so that it has a molecular structure similar to that of petroleum.

8. The second annual ARPA-E Energy Innovation Summit was held during February 28-March 2, 2011, at the Gaylord Convention Center just outside of Washington, DC.

9. George Will, the prominent conservative columnist, frequently inveighs against predictions of global warming, see, e.g. *huffingtonpost.com/2012/07/08/george-will-heat-wave-summer-climate-change-global-warming_n_1657504.html.*

10. A carbon offset is a reduction in emissions of carbon dioxide made in order to compensate for or to offset an emission made elsewhere.

11. Some argue that earth's biosphere may be able to sequestor much of the increased CO_2 in the atmosphere associated with human fossil fuel burning. This effect is known as "CO_2 fertilization." According to this scenario, higher ambient CO_2 concentrations in the atmosphere literally "fertilize" plant growth, while plants in turn, via photosynthesis, convert CO_2 into oxygen.

12. Koch Industries and the billionaire brothers who own it are well known as lobbyists — denying climate change and backing efforts to roll back environmental, labor, and health protections at the state and federal levels. The American Petroleum Institute (API) is also heavily involved in lobbying activity.

13. Methanol, also known as methyl alcohol, wood alcohol, wood naphtha, or wood spirits, is a chemical having the formula CH_3OH (often abbreviated MeOH). It is the simplest alcohol.

14. Methanol-to-gasoline technology (MTG) reactions dehydrate methanol and convert the available carbon and hydrogen into various hydrocarbons.

15. Researchers have attempted to find a way to efficiently convert natural gas directly to usable liquid fuel via gas-to-liquids (GTL) processes since German scientists Fischer and Tropsch successfully converted coal to liquid fuel in the 1920s. It is an energy-intensive process and to date, the number of commercial-sized GTL plants remains limited.

16. Compressed natural gas (CNG) is a fossil fuel substitute for gasoline (petrol), diesel, or propane/liquefied propane gas (LPG). Although its combustion does produce greenhouse gases, it is environmentally cleaner than those fuels.

17. The standard oil barrel of 42 U.S. gallons is used in the United States as a measure of crude oil and other petroleum products. Elsewhere, oil is commonly measure in cubic meters (m3) or

in tons (t). Natural gas is usually measured by volume and is stated in cubic feet. A cubic foot of gas is the amount of gas needed to fill a volume of one cubic foot under set conditions of pressure and temperature. 1 cubic foot = 7.48051948 gallons (U.S. Fluid).

18. A price differential is any difference in the prices charged for the same product to different *market segments* or in different geographic regions.

19. The Renewable Fuel Standard (RFS) program was created under the Energy Policy Act (EPAct) of 2005, establishing the first renewable fuel volume mandate in the United States.

20. Leadership in Environmental Energy and Design (LEED) recognized standard for measuring building sustainability. Achieving LEED certification is the best way for you to demonstrate that your building project is truly "green."

21. In July 2010, the World Bank (Baffes and Haniotis) released a report entitled "Placing the 2006/08 Commodity Price Boom into Perspective" that reexamined some of the evidence. This World Bank report argues that "the effect of biofuels on food prices has not been as large as originally thought."

CHAPTER 5

THE POLITICAL ENVIRONMENT AND U.S. ENERGY SECURITY

Achieving energy security is an important component of U.S. national security strategy. As we develop new approaches to meeting our energy needs and as the global environment changes, this task will become all the more complex. This chapter is based on the remarks of four speakers. Bernard Cole focused on the impact of new energy-hungry powers on U.S. energy security. Robert Cekuta discussed the challenges posed by unconventional threats such as terrorism and international crime. John Bumgarner dealt with cyber attacks, notably on the electric grid. Then, to launch the discussion, Stephen Kelly put the prior remarks into perspective by examining U.S. energy relations with Canada and Mexico.

RISING GREAT POWERS AND COMPETITION FOR ENERGY

Bernard Cole

My remarks will reflect my own views and not those of the National War College or any other agency of the U.S. Government. The organizers posed four questions:

1. How will the rise of new energy-hungry powers affect U.S. energy security?

2. What challenges are posed by unconventional threats such as terrorism and international crime?

3. How will these be affected by a shift toward greater reliance on alternative types of energy?

4. How vulnerable will market grids be to cyber-attacks?

My own recent research has focused first on Asian maritime issues, particularly as affected by the incipient naval arms race in progress in the Western Pacific and Indian Ocean waters, and second on energy security issues in the region. I'll address with particular attention the organizers' first three questions, confident that John Bumgarner will slay the dragon of cyberspace—a new theater not only of war, but of peacetime competition as well.

The basic premise from which I view the international energy situation is that which I've heard repeatedly from experts in the field, both in academia and industry. The premise is that petroleum is a fungible product. That is, a barrel of oil pumped anywhere is a barrel of oil pumped, and it goes into the great pool called the international energy market. That will become important when I start addressing China's current energy policies. I'll briefly discuss Brazil, but the two emerging powers on which I'm going to focus the majority of my remarks are China and India. These are the two countries most likely to affect the U.S. energy security in the future.

Neither China nor India is an enemy or an ally of the United States in the traditional sense of those terms. Both have very strong economic relations with the United States, but neither is close to matching it in terms of economic or military strength. Although a great deal of ink has been and continues to be expended over China's remarkable economic growth over the last quarter century, as well as its far more moderate military modernization efforts, they are usually phrased as potential threats to U.S. allies and interests in East Asia.

The past decade has been marked by two significant developments among China, India, and the United States. First has been the strengthening of the political and security relationship between the United States and India, following September 11, 2011 (9/11). The rapid U.S. military entry into Central Asia, reaffirmation of defense relations with Pakistan, and the notable warming of relations with India also occurred during this period, and all were of concern to China.

Second is China's remarkable economic expansion in terms of both domestic productivity and international trade. However, this leaves many domestic issues of very serious concern to Beijing that we'll have time to address later. The relationship between China and India is both intriguing and problematical, although one certainty lying in the future between these two giants is that their relationship will be of significant strategic importance to the United States. From a security perspective, the general air of antipathy between China and India may bode well or may bode ill for the United States.

Brazil.

Brazil became a net oil exporter very recently, just in 2009. The U.S. Department of Energy (DoE) estimates that this status will continue, but hedges that prediction based on Brazil's own increasing energy consumption. Of interest to the United States is the fact that Brazil's dramatic new petroleum finds are located off shore in very deep water at up to 18,000 feet depth. This means an increased Brazilian interest in the maritime arena and may lead to a modernization of Brazil's naval forces that would be of interest to the United States. Even given the generally disappointing state of affairs between the United States and Latin America, it's unlikely that Brazil's emergence as a major energy exporter, should that develop, would lead to any energy security concerns for the United States. Perhaps far more important is the relationship that is developing between Brazil and China, a relationship that is already experiencing some serious hiccups, particularly on the part of Brazil.

India.

Let me also comment on India's and China's economic expansions and energy security, especially in the sense of ensuring the availability and affordability of supplies. Energy security is an important and sometimes contested issue between China and India. The latter's search for a solution has only marginal energy security implications for the United States. Both India and the United States are intimately concerned with the flow of petroleum supplies from Burma and from Central Asia, albeit mostly from different perspectives.

One source that promises to meet India's increasing energy demands lies in Burma's offshore waters, rich with natural gas. But Beijing has so far emerged the victor in the direct competition with New Delhi for those reserves.

The U.S. attitude toward Burma's repressive military regime precludes any significant cooperation with India over these reserves: this coincidentally formed a major objective of China's National Overseas Oil Corporation's (CNOOC) unsuccessful attempt to buy the U.S. oil company UNOCAL in 2007, an attempt frustrated by the U.S Congress. CNOOC was probably most interested not only in attaining entry into those leases off the Burmese coast that UNOCAL had the rights to, but also UNOCAL's deep sea drilling technology, knowledge which also plays into the Brazilian card because of the deep beds that they had.

Central Asian reserves are also in contention between India and China, but their location makes them difficult for New Delhi to access. Two possible pipeline routes are under discussion; the first is the Tajikistan-Afghanistan-Pakistan-India (TAPI) route, which poses obvious problems: building and operating a pipeline through war-torn Afghanistan is not practical now or in the immediate future, no matter what nation attempts the task. For India, the TAPI route through Pakistan also poses a unique and very serious problem, given the apparently undying enmity between the two countries: what Indian government is going to be able to accept relying for vital energy shipments on Pakistan?

The second oft-discussed pipeline through which India could access Central Asian energy reserves is the Iran-Pakistan-India (IPI) route. This proposal for a natural gas pipeline not only repeats Indian doubts about relying on Pakistan for a vital energy route, but has also suffered from Iran's insistence on pipeline charges, per barrel of product, in excess of what Pakistan is willing to pay for transshipment.

The United States is concerned with these two pipeline proposals on several levels, although probably none of them are direct security concerns in Washington. First is the desire to see India prosper economically—with secure energy supplies—as a friendly, if not allied, nation.

Second is the U.S. concern for the enhanced stability and durability of Pakistan as a nation-state and ally in the campaign against terrorism, a concern intimately tied to the third U.S. concern, that is, our emerging from the conflict in Afghanistan with hardly a shred of credibility for leaving behind a viable polity.

Fourth is Washington's concern with Iran, a political bad actor from the American point of view, with respect both to Tehran's nuclear development program and to its sponsorship of terrorism around the Middle East, including disparate groups in Afghanistan. The United States sees little benefit in the present Iranian regime's gaining in viability through energy profits from India.

A final area of potential energy gain for India that would benefit U.S. security interests is in sub-Saharan Africa, an area in which China has been especially active. India's attempts here have been minimal, compared to China's—although India's investment in Sudanese energy reserves is second only to Beijing's.

India is often lauded in Washington as the world's largest democracy (which is true in a population sense, at least) and itself is undergoing notable economic modernization. In terms of the triangular relationship I am discussing, however, India's real strategic importance to the United States, and the problem it poses to China, is simply its geographic location. For the United States, India may serve as a strategic partner, if not an ally, while for China, India's modernizing navy and capable air force largely counter the strategic effects of Beijing's ongoing acquisition of commercial ports of call that might serve as support facilities for China's modernizing navy. These are generally listed as Mergui or the Cocos Islands in Burma; Chittagong, Bangladesh; Hambantota, on the southwestern coast of Sri Lanka; and Gwadar, Pakistan.[1]

China.

Energy security certainly has become a major strategic priority for China since that country became a net energy importer in 1994. The facet of Beijing's search for that elusive goal that most concerns U.S. strategic interests is Beijing's campaign to "own" petroleum supplies from exploration to consumption. China's international energy companies have—for more than a decade—been following the so-called "Go Out" campaign, meaning they have been involved in foreign energy markets, exploring for and locating energy fields, recovering their products, and shipping them to China for refining or, if refined in the originating country, for consumption.

There are significant qualifications to this paradigm, however. First, China's nationally owned oil companies pumped less than 1 percent of world oil production in 2006. Second, less than one-third of the petroleum obtained in this fashion by China from foreign countries actually reaches China. The other two-thirds are sold on the international market to the highest bidder, either before or during shipment from the originating country to China.[2] Third, of that one-third, no more than 10 percent is carried to China in Chinese-flagged tankers, further denoting the internationalization of the energy trade.[3] Hence, China is expanding the amount oil available globally—to all nations—by virtue of its overseas energy recovery activities.

Beijing's view of energy security is based on three concerns. Its first concern is the availability of energy supplies for China, primarily petroleum products, including those from foreign sources. Its second concern is the affordability of such supplies. Only third—and a distant third, at that— is concern about the military security of energy supplies.

Beijing's determination to secure overseas energy supplies at origin and to increase the availability of Chinese flagged tankers to transport those supplies reflects the first two facets of energy security for Beijing—that is, affordability and availability of energy.[4] The National Resources Development Council in Beijing has stated that by 2050, they want 60 percent of all imported seaborne oil supplies carried in Chinese flag tankers, and they've launched the appropriate shipbuilding campaign to make that happen.

The military aspect of energy security is being demonstrated in part by the modernization of the Chinese navy—albeit very much a work in progress—and in the oft-quoted (or rather misquoted) statements of concern about the "Malacca Dilemma," which refer to fears that a foreign opponent (i.e., the United States) could strangle China's petroleum imports by blockading the Malacca Strait, through which a majority of the nation's trade flows. That is more perception than reality, given the navigational limitations of the Malacca and Singapore Straits, but apparently is nonetheless a factor in Beijing's strategic calculus.

The military facet of energy security is made further questionable by the fact that the strategic reserves of petroleum built by China are stored in above-ground tank farms located on the country's coast. There was absolutely no concern for military security for those tank farms.

Finally, China continues to rely on its very large indigenous coal reserves for about 70 percent of its energy requirements. This is such an inefficiently managed industry in China in terms of both safety and function that should the Chinese government be able to get control of the coal industry to do simple things like washing the coal, allow liquification of the coal at the mine mouth, and increase the efficiency of their transportation network (which is really poor), then they still have a lot of gain from their coal reserves. In fact, Brown and Rupp built a series of coal-fired electric-producing plants in Guangdong Province back in the 1998-99 time frame and found that it was cheaper to import coal from Australia than to bring it down from Northeastern China by train.

In sum, with improved efficiency, China might get a lot from its coal reserves. If I were a Chinese military officer concerned about energy security, the reserve would serve as something of a security blanket. They should temper any draconian strategic decisions as China searches for offshore petroleum sources — decisions that might escalate to military confrontation, an eventuality almost certain to involve the United States.

Unconventional Threats.

Question two, "What challenges are posed by unconventional threats such as terrorism and international crime?" is best answered by "very little," based on past and current events. Much news media attention is drawn to incidents of terrorism at sea, but in fact no more than a half-dozen such incidents have occurred. International crime is a far more serious matter, consisting primarily of human trafficking, exotic animals trade, and narcotics smuggling. These are not going to affect the U.S.'s energy security picture in any serious way.[5]

Human trafficking is estimated to involve 1.4 million persons and yield $31.6 billion annually.[6] The illicit narcotics trade is estimated to amount to an annual $88 billion by the United Nations (UN), but that is, by definition, an approximate figure.[7]

Another form of international crime is widely reported as "piracy," although almost all such incidents are more accurately called robberies at sea, since they do not occur in international waters. This has not posed a serious problem despite all the press. There are two reasons for saying this. These incidents are estimated to cost international trade between $2 and $3 billion dollars annually, but even that sum amounts to less than 1 percent of that trade's value.[8] Right now, from a merchant ship owner's perspective, it's cheaper to pay ransom to the pirates than it is to spend the money on the necessary security procedures to prevent the piracy or robbery from occurring in the first place.

Second, if we look back at the history, the only antipiracy efforts that have been successful going all the way back to the 17th century in the Caribbean, have been those that penetrate to the beach and destroy the pirate's home base. Given the situation in Somalia today, that's not likely to happen in the near future, but that is going to be the only final solution if there is a final solution, which there won't be from a global perspective since piracy has always been with us.

Alternative Energy.

The conference organizers' third question, "How will these be affected by a shift toward greater reliance on alternative types of energy?" is more difficult to assess than are the first two. China currently is a world leader in developing alternative forms of energy, by which I mean non-fossil fuel sources.

Hence, I categorize nuclear power as an "alternative type" of energy, despite the waste disposal problems that it engenders. If China actually builds and brings into operation all the 50 or so nuclear reactors discussed in terms of 2050, however, they will produce no more than 6 percent of the nation's daily electricity requirements—so Beijing estimates. Other forms of alternative energy being developed in China, from biomass to wind power, will certainly improve the country's level of energy security, but will have a marginal impact on the continued reliance on fossil fuels, especially the coal currently providing approximately 70 percent of daily energy needs.

India is also pursuing alternative forms of energy, as are Japan, Taiwan, South Korea, and Australia, but none of these programs may currently be expected to contribute to a dramatic reduction in reliance on fossil fuels unless those two famous—or infamous—researchers out in Berkeley actually discover cold fusion in a jar in their garage.

Japan and Taiwan are particularly bereft of indigenous energy resources. Japan, therefore, is a world leader in use of nuclear power and in the future is likely to increasingly turn to natural gas (including the liquefied product) particularly from the Sokoline fields. Australia is not likely to substantially reduce dependence in its indigenous energy fields, while Taiwan and South Korea appear destined to continue relying almost wholly on energy imports.

Environmental factors are worth bearing in mind. As of 2000, Taiwan had a fourth nuclear power plant under construction. The Democratic Progressive Party (DPP) government that came into power in June 2000 was very much a green party, and they initially tried to stop construction of that fourth nuclear power plant. They had to reverse that decision, but still this is a demonstration of environmental concerns where increased use of nuclear power is involved.

Environmental concerns have also been shown by China. I mentioned earlier their large projects for liquification of coal at the mine mouth. These were stopped by Beijing at the direction of the Chinese government in 2007 because of the concern for the fresh water usage entailed. In other words, fresh water as a resource was more important to Beijing, and remains more important to Beijing today, than the availability of more efficient coal reserves or more efficient energy from coal.

Alternative energy programs should not be expected to challenge U.S. strategic interests, although if the definition of an "energy-hungry great power" is stretched to include North Korea and Iran, their programs apparently in pursuit of the dual use of nuclear technology certainly are of concern.

China is the nation of most concern to American policymakers, having possessed nuclear weapons for nearly 50 years and conducted its first nuclear detonation in 1964. While that does not reduce U.S. concern about an increased Chinese nuclear arsenal, there presently is little indication that Beijing is planning to abandon its apparent strategy of minimal nuclear deterrence. When General Chang Guang Kai, who was then their head of intelligence, had a meeting with retired U.S. Ambassador Chazz Freeman many years ago, he said in so many words that as long as we can put one nuclear warhead on Los Angeles, that was enough to deter the United States. And he was probably right.

India has been a nuclear weapons power for a much shorter period, detonating its first weapon in 1998. A notable indication of Washington's desire to form a close strategic relationship with India was the 2008 decision to acquiesce in that country's status as a nuclear power, despite New Delhi's refusal to adhere to the nonproliferation treaty.[9]

Nuclear power as an alternative source of energy poses dual use dangers to the United States (and to other nations), but under proper engineering and managerial control is a welcome alternative to fossil fuels. This is especially true in view of the major powers' possession of large coal reserves with that fuel's current environmental drawbacks.[10] That's my own view. Having suffered through 16 weeks at the Navy reactor 50 miles outside of Idaho Falls over the winter, I still remain a fan of nuclear power.

As noted earlier, I will leave the conference organizers' fourth question, "How vulnerable will market grids be to cyber-attacks?" to my colleague, Mr. Bumgarner. From a security viewpoint, cyber is now recognized by U.S. military commanders and civilian leaders as a new theater demanding national security policymaking and strategy. The Pentagon is now organized to deal with cyber issues, with one emerging great power—China—evidently the primary country of concern, as time after time hostile incidents of cyber intrusion are traced to that country.

CONCLUSION

China and India are the world's most significant emerging nations—in terms of population, expanding economies, and modernizing militaries. Their position vis-à-vis U.S. strategic security is problematic but mutable; the ability of the United States to affect the paths followed by these two Asian giants is quite limited. Neither is likely to pose an actual threat to the United States, but both are certainly challenging the U.S.'s post-Cold War hegemonic position in the world.

UNCONVENTIONAL THREATS TO ENERGY SUPPLIES

Robert F. Cekuta

I shall address unconventional threats to energy security and energy stability. We at State focused a lot in the 1970s and 1980s on oil. Oil became a less prominent issue in the 1990s and the early 2000s, but now given questions arising due to the situation in the Middle East, many would say oil's back. The reality, however, is that the issue never really went away because one of the basic pieces of our own national security and prosperity is the access to energy.

Terrorism.

The first such threat that comes to mind is terrorism. Iraq offers a good example of the kind of effect terrorism can have. We saw — and still see today — insecurity in parts of the country. According to press reports, bombings and other forms of violence add approximately 10 percent to contract costs. Companies are forced to keep their personnel in secure locations in Baghdad and Basra rather than letting them live closer to the field, in places where they might be better and more easily able to develop the country's energy resources. The attacks on the oil refinery recently and the attacks on power transmission lines, pipelines, and other similar targets that took place in the earlier days of the conflict, did have an effect. The attacks on the energy infrastructure significantly slowed the country's development, and prevented it from moving forward in the period after the fall of Saddam in the way that we had hoped.

But the damage, while insidious, especially psychologically, actually had a rather limited physical impact. You can repair pipelines in relatively few hours. You can usually repair a pumping station in 2 or 3 days. Of course, you've got to get to it, which may not always be easy. But the point here is that the damage that can be done to one of these facilities is actually rather finite, easily contained, and can be fixed. That certainly doesn't mean you can ignore it, however.

Another part of the world where terrorism has been a factor is the Niger Delta. The unrest in that part of the world, especially in that part of Nigeria, can significantly affect production. It results in the loss of something close to 100,000 barrels a day, according to some experts. Those of us in the U.S. Government frequently talk to the Nigerians about what might be done to help improve their internal situation. We talk of the need to boost respect for human rights, to pay attention to the environment, and to make progress on good governance. We also talk of the need to improve transparency so as to build good governance and to give the people a sense that they are empowered and are involved in building their country.

Interestingly, oil companies are doing the same thing. When you talk to the major oil companies operating in the Niger Delta, they will talk about what they are doing to increase employment in the region, to hire subcontractors, to involve local populations in the industry. This is important not only because it will make the area more prosperous. It is also politically important because it makes people see the system as something that they have a stake in and see the oil industry as something that doesn't just belong to outsiders, but as something that is important to Nigerians and benefits them.

Piracy.

The second unconventional threat that is frequently in the news of late is piracy. However, if you have a sense of history, you realize that the problem is not new. People have been concerned for years about the safety of shipping through certain chokepoints. The most famous of these are the Straits of Hormuz and also the Straits of Malacca. In Malacca, piracy has been a problem for years. The governments there, however, are quite strong. They're able to take action, and they've had help from outside. So yes, piracy is a problem here, but it's contained.

Piracy in the nearly one million square mile area off the coast of Somalia has been much more in the news and has proven something that is much much more difficult to handle. As Professor Cole noted, the cause of the problem is a state that many describe as failed. Somalia has been in a state of unrest for over 20 years. It has been hard for any leader to really assert the needed control. Pirates can therefore work with considerable impunity from a large geographic base. Somalia is a large country and has a very long coastline. Moreover, the pirate attacks are not necessarily occurring just off the Somali coast, but well out into the Indian Ocean.

There is something else that needs to be noted: piracy is a business. We had a conference this past week in the State Department where we looked at the questions of how to trace the money flows that go to the pirates. Pirates are getting ransom payments — the money they receive must go somewhere and get into the financial system. If, as some experts think, there are indeed higher-level people who direct these piracy operations, who work with contacts in Somalia who in turn work with the tactical operators who actually get in the boats and go out to sea, we need to identify these higher-ups and figure out how to stop them. How do you find out who they are? You look at where the money goes. How do you stop them? That isn't always easy, as we know from our efforts to track money flows from narco-trafficking, but it can be done. You can find out who the people are by following the money and taking action. Still, as we know from fighting other criminal networks, even when you know who the bad people are, it doesn't necessarily mean it is possible to get them right away.

Yes, piracy is real. It's a problem. Recently pirates hijacked two tankers carrying over $200 million in oil, including one that was en route to the United States. But it really has not affected the flow of oil or upped the price of oil. It has not really damaged our energy security. It certainly is a problem and needs to be confronted, which we are doing. But as a threat to our energy security, not many would argue that the activities of these pirates are especially significant.

Poor Governance.

There is another set of factors—by no means a threat, which can affect energy supplies and security. These are the policies of governments themselves. The economic and political policies that are put in place in different countries affect which companies go in as well as what is done by way of drilling, exploring, identifying, developing, and producing energy. Iraq again offers an instructive case. Iraq may be the world's second-largest oil province. The Iraqis have long had a very conservative way of estimating their reserves. That was true even back in the days of Saddam Hussein. These attitudes and methods meant they were relatively restrained in terms of what they announced in terms of the size of the country's oil reserves and potential production yields.

Looking at the situation now, Iraq's exports over the last 2 months have averaged about 2.7 million barrels a day. That's an increase of 8.2 percent of exports from the 2010 average of 1.9 million barrels a day. Iraq's average selling price for its crude oil exports was about $97 a barrel in February. That's up from a January average of $90.78 a barrel, $86.31 in December, and $80.59 in November, although the rise in world oil prices is a key factor here. Still, revenue from the southern oil center, Basra, climbed to about $4.8 billion—that's the highest since 2003—from exports in January of 54 million barrels. The exports from Iraq's northern oil fields, which include the crude from the Kurdish regions, rose to 494,000 barrels a day. That figure also includes about 10,000 barrels a day taken by truck to Jordan.

Those are rather significant amounts of oil. The Iraqis are talking about trying to increase capacity by another five million barrels a day. They very well could be able to do that over time, depending on what size their reserves really are, but first of all they face the political challenge of instability. Second, they face the challenge of working out a new hydrocarbon law that actually sets up a national system. This internal legal/political situation has been confronting the Iraqi government for years, an issue which needs to be worked out between Kurdish areas in the north and the capital in Baghdad.

There are the infrastructure problems facing Iraq as well. Pipelines that have been around for a long time get old; they need to be replaced. Pumping stations and other infrastructure have also aged and need to replaced, upgraded, or otherwise modernized. The capacity of Iraqi ports is also limited. Revenues from existing oil reserves could likely pay for these improvements, but it's going to take time.

We see the same sort of discussions regarding the legal situation and its potential impact on investment and energy production in other countries as well. Company representatives come in and ask about Nigeria and its new hydrocarbon law, noting that whatever law is put in place must make it easy for the oil companies to work with the authorities there and enable all those involved in developing and producing hydrocarbons in the country to prosper further. I've had this same conversation with companies about what needs to be done in other countries, including new oil provinces — places like Ghana and Uganda that are coming online. Governments in these countries are trying to figure out how best to capture and utilize revenues from new oil finds, how to make sure that they're being used for good purposes and helping national development. They have seen the corruption that has occurred in other places and want to prevent it. They recognize that increasing the transparency of operations and money flows can help them in this effort.

In sum, failure on the part of governments in oil-producing countries to deal effectively with these economic and political issues, while not exactly a threat, can be a factor in limiting the global supply of crude.

The Global Market and Energy Security.

The global market today really has changed significantly. As we look at oil prices, at economic statements in the press, and at concerns regarding Libya, it's clear that we are not living in 1979. There may be reports of instability in the Middle East, but we're not looking at gasoline lines in the United States again. That's because of the work we have done since 1979.

First, we've built up Strategic Petroleum Reserves roughly 146 days of normal consumption are sitting in cavern and tank facilities around the world. That is an important factor in keeping prices under control and ensuring our economic well being.

Second, we are better about engaging in discussions with producers than we once were. I was in Riyadh recently for the 20th meeting of the International Energy Forum. This is a group that brings together Organization of Petroleum Exporting Countries (OPEC), countries belonging to the International Energy Agency (IEA) (which was set up in 1973 after the Organization of Arab Petroleum Exporting Countries [OAPEC] boycott of crude to the United States). These are the major energy-consuming countries and third world countries. The group looks at what the situation is in the oil market, looks at demand, and looks at production levels. It doesn't try to fix prices—you can't do that, the interplay of supply and demand in the marketplace can do that. But the associated International Energy Fund (IEF) can help make sure that there's a flow of accurate information so that producers know what to expect by way of demand, and consumers understand what to expect by way of production.

A number of statements by Ali al-Naimi, the Saudi Oil Minister, and others came out of that meeting. The

Saudis expressed a willingness to put crude oil on the market. There were conversations subsequent to that conference between the IEA and the other industrialized countries. Statements from the IEA at the same time pushed oil prices down $6.00 a barrel. We have also talked to oil companies. They say there is no shortage of crude. The Saudis, with at least a couple million barrels a day of excess capacity, said they would make crude available to anybody who wanted it. Thus the oil seems to be there right now in the physical market.

The Future.

There is another big factor which comes up when we talk about the stability and security questions. We talked about oil supply and demand 20-30 years ago among the Organisation for Economic Co-operation and Development (OECD) countries, such as the United States, France, Austria, Germany, and OPEC. Today, however, you see our demand in the OECD countries starting to decrease, but the demand of China and India going up sharply. There are parts of this world that really have not developed at all but are going to have to—places like Africa, which has 800 million-plus people. Right now the total energy consumption in Africa, according to the IEA, is equal to the energy consumption of New York State. The consumption and demand numbers have got to rise. However, there is still the question of *how* these countries are going to develop? Are they going to develop like us or not? These are factors we need to look at as we go forward.

There are two other important factors on the supply side that affect today's energy equation. First of all, consider the great boost in natural gas production in the United States and the tremendous impact this has had in terms of our energy security and energy supplies. Then there are the steps we are taking to encourage the development of low-carbon technologies and of renewable energy technologies.

These are matters which we at State discuss with the producing countries. They're aware of these developments, and they're concerned about them in terms of global market stability. Such discussions are important for all.

CONCLUSION

In sum, there are indeed unconventional threats. The factors that keep us awake at night, however, are the increased demand in the rest of the world and the effects this is going to have on us, particularly short-term shocks. At the same time, we have confidence in the steps we have already taken that can help us ameliorate those short-term shocks.

SMART GRID VULNERABILITIES TO CYBER ATTACKS

John Bumgarner

INTRODUCTION

When I entered the security field nearly 3 decades ago, most people didn't have a computer, let alone an Internet connection. Now everyone in this room has a least one computing device and uses the Internet for both work and play. Unfortunately, our national security efforts haven't kept up with the threats that have accompanied these advancements.

For example, America's electrical grid and other critical infrastructures are vulnerable to cyber attack from other nation and hacktivists. Threats will only increase unless the DoE and the Department of Homeland Security (DHS) take major steps in the coming years to reduce our nation's cyber vulnerabilities in the grid.

The U.S. electric grid is undergoing a transformation which will make it smarter and more efficient. These modifications will take years, if not decades, to accomplish, however. As an industry we need to engineer these newer systems to be more secure from cyber attacks or technological mayhem. Unfortunately, the electric industry is deploying technology with common cyber vulnerabilities embedded.

DAMS, TURBINES, AND GENERATORS

In 1882, when the Pearl Street Power Plant was completed, computers didn't exist in the world. When the Hoover Dam was built during the Great Depression, America wasn't worried about cyber threats. The Francis turbines installed at the dam weigh in at approximately 172 tons.[11] Today these gargantuan turbines are computerized, but they were never engineered to withstand a cyber attack. If a cyber attacker damaged the turbines in the Hoover or Grand Cooley dams, America would have major problems.

The first problem would be the immediate loss of electrical capacity for the dams' electric customers. The second would be replacing the damaged turbines, which is extremely problematic because America doesn't manufacture them. These turbines are made overseas and designed to precise specifications for each client. How fast could American replace damaged turbines inside the Grand Cooley Dam?

Engineering flaws also exist in coal-fired and gas-fired generators, which can be exploited by cyber means. Unfortunately, prospects get worse, because nuclear reactor cooling systems are also vulnerable to cyber manipulation. A coolant pump installed at Iran's Bushehr Nuclear Power was damaged when it was started. This coolant pump was connected to a programmable logic controller, which was controlled by a computer. What do gas centrifuges and coolant pumps have in common? The answer is "rotation." Generators, turbines, centrifuges, and coolant pumps are very unforgiving when it comes to rotation disruption. Catastrophic damage usually occurs to key moving parts in these systems.

There is a subset of individuals' within the DHS, the Department of Defense (DoD), and the intelligence community who are highly aware of these types of cyber threats. Many of the organizations know that hostile individuals are actively mapping vulnerabilities in American's critical infrastructures.

Many of the vulnerabilities embedded in deployed generators and turbines are not going to disappear anytime soon. What we need to start thinking about is reengineering critical components currently on the drawing board. Some people say that it's not cost effective to do this because cyber threats are not a major concern.

Unfortunately, Idaho National Labs has already proven this statement false. A few years ago, the lab conducted a security assessment of a diesel-powered generator. The purpose of this test was to determine if a cyber attack could damage the generator.

We have a video of a diesel-powered generator being attacked. A viewer cannot see the technicians manipulating the generator rotation mechanism to cause excessive vibrations, just as an actual observer could not see a hacker. The generator was never engineered to handle this level of vibration. The hacked generator does not like being manipulated and broke apart. How do we prevent such intentional attacks in industries across America? Today, we cannot, but in the future we will protect them.

Unfortunately, America does not manufacture the majority of the generators installed in the country's electric grid. So how does the State Department ask a foreign power such as China to put America at the head of the manufacturing line for a new generator? As you can see, this can quickly become a major foreign policy problem for our government.

One of the past research projects of the U.S. Cyber Consequences Unit (USCCU) was to identify the production cycles of critical components in the electric grid. We discovered that the normal business cycle for producing a standard generator was about 18 months to 2 years. Some of the more specialized components take 5 years. That's a long time to wait, but America is competing with all the other customers of the world. The United States has to consider establishing relationships with Canada and Mexico for sharing the costs of maintaining critical spares. These relationships could possibility include formal agreements.

HOME ELECTRICITY

Now consider another scenario. Everyone here probably lives in some type of house or apartment. Everyone's probably got an electric meter on the side of his/her house. Most of the electric meters installed today are like those designed in 1887. Over the last several years, only a small fraction of the meters in

America have been upgraded to smart meters. These smart meters are one of the pieces of a smarter electrical grid that is being installed in America and other advanced countries. We are witnessing an evolution in the utility industry, which will eventually impact the three primary utilities of electric, water, and gas. Meters in these industries will be replaced with ones that are computerized and networked together. This is a multi-decade project, so we need to start thinking about the security of the systems before we deploy hundreds of millions of the meters across the United States.

Most companies that have deployed the meters have focused on reducing business costs and not security controls. For example, public records of Florida Power & Light suggest that the rapid disconnect and reconnect was a primary reason for investing in smart meters. These meters also reduce workforce costs, because fewer employees had to read meters or make service calls to disconnect or reconnect power at homes. Computerization made all this possible from a remote location, but it also opened the door for hackers to access the systems from remote locations.

This move to the smart grid, moreover, is not just taking place in America. There is a new market for prepaid electricity services. In the Pacific Islands, they are putting prepaid meters in their homes. To get services, you have to go inside and enter your code that verifies that you have electricity, and that you're paying for it.

Vulnerabilities of the Smart Grids.

Why is the use of smart grids a cause for concern? One major problem facing the industry today is the lack of government security standards. The National

Institute of Standards in Technology has been developing standards for the electric industry, but not for the other utilities.

Another problem is the large number of meters that are going to be deployed. According to one estimate, there were 80 million single-family homes in America in 2010. That's 80 million electrical meters. That doesn't take into consideration businesses, apartment complexes, and "mobile" residences. We're talking hundreds of millions of these devices that have to be deployed. That's hundreds of millions of computers. If we don't take due diligence today to design them with security built in, someone will hack them.

Last year, I wrote about a future threat posed to smart meters from something called "warmetering." This activity is when hackers with specialized equipment and software ride through neighborhoods looking for smart meters. The hackers use this information to map meter types, which can be useful in identifying vulnerabilities.

Some people in the utility industry think that this type of activity is unlikely. Last year at the DEF CON® Hacking Conference in Las Vegas, a security professional demonstrated how to hack a smart meter used in the electric industry. The hacker was able to modify the liquid crystal display (LCD) to read "pwned," which is geek-speak for owned.[12] The hacker also displayed a computer worm that could compromise the smart meter remotely. In a simulation, the hacker used geo-location information to program the worm to target the city of Seattle. The attack simulation had the worm spreading across Seattle area, disrupting smart meters it encountered. Within minutes, the worm had infected 200,000+ meters.

Recently, I was talking to students about integrity, availability, and privacy issues of smart meters. Part of our discussion centered on cyber attack response by utilities. I explained that a weather-associated event, such as a power line going down, is usually an easy repair. I also explained that utility workers could easily replace a few damaged meters. But what if 300,000 meters went offline at the same time? How would the utility respond? If the meters are damaged, would the utility have enough spares? If the utility does not have spares, can they acquire them from other sources? What level of costs would be associated with such an incident?

Another concern, which hasn't been fully addressed, is the environmental impact associated with replacing hundreds of millions of old meters. The old meters being replaced contain harmful metallic components — some contain silver, some have toxic metals. How do we recycle these old meters without damaging the environment? Will the Environmental Protection Agency (EPA) establish recycling guidelines for smart meters?

Green technologies also have security problems that need to be addressed. For example, wind turbines have security vulnerabilities, but most companies are not securing them from cyber attacks. Wave generation equipment and solar arrays also have surfaces vulnerable to cyber attack.

Even electric cars have potential vulnerabilities that hackers could exploit. We need to start looking at the cyber attack surfaces in cars. Hackers are already testing the security of cars. Imagine a mischievous individual driving by a car lot and ordering all the cars to flash their lights and beep their horns at the same time. Electric cars will also be embedded with chips that will allow utilities companies and the government to track your mileage and location each time you charge up.

The other thing to think about is power infrastructure. Google and Microsoft are already thinking about it. Google's doing a pilot project with smart meters. Originally, it involved only one company, but now it involves multiple companies—probably well over a dozen. Google is developing this technology called Power Meter, which is connected to the Internet. You can plug a device into your home that will talk to your meter, and then it will send that information to Google. Google collects all this information and compiles it. Using this information, Google can send you a coupon to purchase a more efficient clothes dryer.

Utilities will have to work out ways to integrate more efficient systems into consumers' homes. Accomplishing this level of integration will take years. In the future, consumers will have their Heating, Ventilation, and Air Conditioning (HVAC) systems; water heaters; and certain household appliances connected to the power grid for the sake of efficiency.[13] These devices will be able to be remotely controlled by the utility and the consumer. Hackers will be looking for vulnerabilities in these networked and computerized appliances. Imagine your dryer bursting into flames because some hacker launched a cyber attack. It's already technically possible to start a fire by launching a cyber attack on a smart outlet. Will arsonists burn down homes and businesses in the future using computer code?

Another area that I'm concerned with is privacy. Last year, I coined the term Personal User Information (PUI). In the future, utilities will be able to monitor movements in your home. When you get up in the morning and turn on your coffee pot or use your washer and dryer, the utility company will be able to see what time you get up and where you go in your home.

A final area of concern is how to protect the power grid from cyber attacks. The DHS first became interested in cyber attacks on the electric grid because it was concerned about what these would do to our economy. The USCCU analysis showed that a regional cyber attack on a city can be a big problem. The impact can vary, depending on how much damage is done and other factors such as weather. Imagine a cyber attacker disrupting power in North Carolina during the summer months when we depend on air conditioning. If you can disrupt electrical power for 7 or 10 days in a given region, you can cause a 70 percent loss in economic productivity. Some of our attack scenarios focus on disrupting electrical power for 3 to 6 months. Could the city of Raleigh or Charlotte, North Carolina, or Washington, DC, survive a power outage for 3 to 6 months? No.

COMMENTARY

Stephen Kelly

As far as national threats are concerned, we've got problems. In no particular order, we've got China and India as potential strategic competitors. Most of the world's consumption of oil over the next 20 to 30 years will come from China. China has already become the largest emitter of greenhouse gases, and the Chinese are well on the way to becoming the largest consumers of oil in the world. Regardless of what you think about peak oil, it's probably a finite resource. The Chinese are quite aggressive in pursuing it. The Chinese don't obtain it for their own use necessarily, they're out there playing in the markets, with implications for us. Then there is Brazil, a net oil exporter starting in 2009, but with growing domestic consumption.

Reports on terrorism and piracy and the effect they have on oil production in Iraq and Nigeria are sobering. We've learned how political policies limit supplies of crude because they sometimes result in an inefficient exploitation of resources. Then there's the possibility of cyber attack.

I'd also like to broach Canada and Mexico. These are areas of the world that we didn't much touch on here. Canada in particular is one region we might want to discuss in the oil context specifically, but in the energy context more broadly.

In 2010, according to the U.S. Energy Information Administration (EIA), the United States actually imported about half the oil it needed on a daily basis. That's down from about 60 percent as recently as 2005. The recession reduced the amount of oil we needed overall. But thanks to offshore drilling, we're actually gradually producing more oil domestically. So much for peak oil in the United States.

Of the oil that we have to import (roughly 10 million barrels a day), 2.5 million of those barrels came from Canada, mostly from the oil sands in Alberta. All of that came to us in pipelines. We've discussed pipelines today as another point of vulnerability, and indeed the pipeline network in North America is not perfect. There was a pipeline leakage into the Kalamazoo River in Michigan this past summer. There are a couple of gas pipelines that keep blowing up. So pipelines are not perfect. Nonetheless, oil that comes to us in pipelines is not going to spill the way that the Exxon Valdez did—there was a story in the paper recently about the continuing effects of the Exxon Valdez disaster in Alaska. Pipelines will not lead to another Deep Water Horizon disaster.

Canada actually provides 22 to 25 percent of the oil we need to import. Of course, one of the big issues of the day is the proposed construction of a new pipeline from Alberta to Port Arthur, Texas. This would bring an additional 400,000 barrels a day of Canadian crude to Houston where refineries exist that can handle the Canadian crude, which tends to be kind of heavy and less tractable.

A barrel of West Texas Intermediate (WTI) closed yesterday at $101.91.[14] But that's just a benchmark price. But a barrel of Brent—North Sea Oil, also a benchmark—closed at $114.79.[15] There has been about a $14.00 to $15.00 difference in what oil costs in the American southwest and what oil costs in the North Sea for several weeks now. Deutsche Bank predicts that this difference will continue to exist for some time.

Thus, Trans-Mountain Pipeline, Inc., (TMI) oil tends to pile up, lacking an outlet, so this is somewhat of an artificial situation. But it does point toward a need to reevaluate potential solutions for some of the energy security problems that have been identified today. There aren't a lot of terrorist groups in Canada. The country's had a few, but it's been a long time since they were active. The pipeline network is actually pretty good. If the Keystone pipeline is built, obviously it would be state of the art. Canada is, of course, democratic, has a good human rights record, and is also a good neighbor, so you don't have a lot of cost in transportation. Relying on Canadian oil seems like a smart option.

Of course, the Canadians' protection of their own grid against cyber attack may also be problematic. In the northeast of the United States in August 2003, there was a great power blackout that cannot be attributed to a cyber attack. A low-hanging wire in Ohio touched a tree and shorted out. The resulting series of failures in the system led to a blackout that covered eight U.S. states, the province of Ontario, and 50 million people.

The grid is thus already fairly vulnerable. One can also go back to 2001, the great electricity crisis in California. That was mostly a problem of lack of infrastructure and capacity to move electricity around, perhaps resulting from market forces. Mexico is our number two oil supplier despite the many problems they have in their industry. Last year, they accounted for 11 percent of the oil the United States had to import. Between Canada and Mexico, that means more than a third of our decreasing percentage of imported petroleum came from our immediate neighbors.

We produced domestically about 90 percent of the natural gas we need. Ninety percent of the 10 percent that we currently import also comes from Canada in pipelines. With the shale gas coming online, we may become self-sufficient in natural gas, which means that energy security in the broadest sense comes down to oil for the immediate future.

QUESTIONS AND ANSWERS

Q: Dr. Cole, you mentioned the China/Indian relationship. Could you comment more on what's driving tensions there? Is it purely a competition for resources? Is it just a question of unfriendly neighbors?

Bernard Cole: Beijing does not consider India to be any sort of conventional military threat. The tension on Beijing's part is not because of India's conventional military forces. The Himalayas are an impenetrable barrier, but rather because India is a nuclear power. It's especially concerned with respect to India's relation to Pakistan, which China considers to be its formal ally. There is one China-India territorial dispute over the boundary that was drawn by the British in 1906.

But India does have security concerns about China rather than the other way around. China is feeling very confident in economic energy terms right now, and in the one or two cases where the two have gone head-to-head in terms of securing energy deposits—Burma is a case in point—China clearly has emerged victorious over India.

Q: This is just a brief comment about Mexico. In March 2011 at the Milken Global Conference, President Vicente Fox said that the Mexican government has used Petróleos Mexicanos (PEMEX) as a cash cow and not really reinvested in it, and has thoroughly abused Mexico's oil wells. He said that because of this, Mexico is on track to stop exporting oil a few years down the line unless they drastically change their rules regarding foreign investment and so forth.[16]

Stephen Kelly: That's absolutely true. The Mexican constitution prohibits outside investment—foreign investment in the upstream energy sector, and actually the downstream one, too. The word privatization is a swear word in Mexico when it comes to the oil industry. As to PEMEX, it was nationalized in 1938, when U.S. and British interests were kicked out. Most Mexicans are very proud of the fact that they kicked out the British and the Americans.

But there are some changes. President Calderon managed to get through some energy reforms 2 years ago, although they were a bit watered down.[17] In the recent meeting between President Obama and President Calderon in Washington, in addition to resolving a long-simmering trucking dispute, there was mention of a cooperative energy development. There are a number of oil reserves in the Gulf of Mexico that are probably transboundary, and the State Department has been working on ways we could exploit those transboundary reserves together. It would be very damaging if both countries dug in — drilled into the same oil field and pumped separately, so doing it cooperatively is in both of our interests.

Mexican oil production has been declining for several years. Their largest field, Cantarell, probably peaked in 2003.[18] The new deposits that they have found have not come online as they expected. They don't have the capital to do as much deep water drilling as they'd like to because, in fact, PEMEX is used as a cash cow for the rest of the government.

But the Mexican government fully recognizes these realities, and it's actually in our strategic interest as a country to help Mexico reverse that situation. We're probably not going to be able to do it by urging them to change the constitution because we were the object of that change back in 1938, but there might be other ways that we could work with them to stem this decline in production. We have an interest in so doing.

Robert Cekuta: A couple of things to bear in mind. One is that the Mexican government, and PEMEX too, has been taking steps to open up more to the outside world. I think part of this becomes a question of how you handle something politically when it's as important to the national identity as the notion of having kicked out the Americans and the British. You can't just turn back on that. We can think of aspects of culture that would be extremely difficult to reverse. But we are seeing the Mexicans become much more interested in talking with other countries and with companies outside Mexico.

The second thing is that we do have ongoing talks right now. We started in December trying to deal with transboundary questions in the Gulf of Mexico. These talks are now underway. I'd just leave it there in terms of talking about the discussions with Mexico.

The lesson of the Arabian Oil Company (Aramco), a giant four-company combine, is instructive.[19] The Saudis nationalized Aramco. They've treated the combine effectively as contractors or subcontractors, keeping Aramco at arm's length from the Saudi government. The Saudis are very proud of the progress that Aramco and the country have made in terms of exploration, exports, efficiencies, good field management, and so forth. They've used the combine, even though Aramco has been nationalized, to get the expertise they need. It's been a relationship that both sides have profited from.

Abu Dhabi is similar. The Abu Dhabi National Oil Company is very open to the notion of working with the international oil companies and taking advantage of their expertise. They are looking into how this might benefit them (and us) by getting more oil out of the ground and into the global market.

What about the countries that have not done this? When I was posted in Iraq in the mid-1980s, I spoke about this with a person I knew well who had a leading role in the Oil Ministry. He said that the Iraqis, under the influence of Arab nationalist ideas, did not want anything from those nasty Westerners. They turned away and either developed technologies themselves or got it from other non-Western sources. He said there were seven engines on a Soviet-designed oil well, but only one was needed on an American well. That's the difference in levels of efficiency. Reports out of Iran speak of the lower productivity and the declining reserve figures, much of it resulting from the Iranians trying to do it themselves and consequently often mismanaging the fields.

The experience is similar in other countries. If you don't get the engineering right, then you damage fields, and you can't fix them. Thus the point about PEMEX is a good one. The Mexicans are making changes there, taking steps, doing it quietly, doing what's important. There are other countries that face a similar situation. We can find examples of countries that haven't done it right and point to the damage that results.

Q: I have a question for Dr. Cole, but first a comment. Aramco is within arm's length from the government. I suspect that they take technology where they can get it. In my experience, they exploit new Western technology faster than any other company, including Shell and certainly faster than Exxon. On the other hand, the influence of the government is very strong, make no mistake. Aramco is allowed to do only what the Saudis want.

I've always wondered what CNOOC was doing in trying to buy UNOCAL. It doesn't matter who owns oil, it goes into the global pool. Very few drops actually go back to China. You indicated that one of the reasons was to get deep water technology from UNOCAL and that's correct. China does have the technology and exploits Brazil, you said. You also said that the Chinese were developing relations with Brazil better than anyone else. But if they are exploiting Brazil, I would question whether they're going to operate in Brazil.

Bernard Cole: This is an important point, because after President Hu Jintao made his Latin American swing several years ago, the press was full of reports about the promised $100 billion Chinese investment throughout Latin America. That simply hasn't developed. Now, especially since the recent change in government in Brazil, there's a lot of resentment among Brazilians as they come to realize that the macro deal they thought they were making with China hasn't come into force. The Chinese are happy to sell their products in Brazil, but their purchase of Brazilian commodities, primarily soy and potentially oil, has simply not developed the way that initial press reports indicated.

Stephen Kelly: CNOOC just purchased a portion of the Chesapeake Oil shale gas field in Texas. Considering that shale gas is the wave of the future, the company is certainly active. Even bearing in mind that 77 percent of the world's conventional oil reserves are in the hands of national oil companies, CNOOC is a special case. It's 65 percent owned by the government, and the other 35 percent is owned by more than 400 different entities, no one of which has more than about 1 percent of the total. It's an interesting creature. The Chinese have gotten a foothold in a very important part of the American shale gas market because Chesapeake, of course, is the largest of the shale gas companies.

Audience: What is the objective of doing that? What do they get out of it?

Stephen Kelly: They probably get technical knowledge. Chesapeake's a leader in horizontal drilling and fracking. There are shale gas deposits in China as well, but then again energy's a worldwide business. The United States may become an exporter of natural gas if the environmental concerns around fracking are satisfactorily dealt with. Perhaps it's just another investment.

Comment: One quick comment on the UNOCAL purchase. This came out during the hearings, and it may or may not have been a primary purpose of CNOOC. UNOCAL actually owned Mountain Pass, which is the only rare earth mine in the United States (though not operational). Since China essentially produces 100 percent of the rare earths that we consume and since rare earths are absolutely critical for defense applications, there was a lot of agitation over that.

Robert Cekuta: Molycorp is looking at reopening Mountain Pass right now. Linex is also doing mines in Australia. So the situation was changed a little bit.

Audience: This question is for Mr. Bumgarner. Could you comment on the recent cyber attack on Iranian nuclear facilities? If it was someone from the West, doesn't that invite a counterattack on our facilities?

John Bumgarner: The worm you're referring to had to be written by a nation-state. It was said that it took 6 months to develop, but that's incorrect. Stuxnet took years to develop.[20]

The original code was developed many years ago, and it was then modified and released. There's some question right now about where the original infection occurred. Someone says they found an infection in Australia in June 2009, but no evidence has been made public to support that claim. Someone else said they found evidence in Mexico in July 2009. But again, they still don't have any technical evidence to support those claims. The evidence out there shows that whoever developed this knew what they were targeting and how to go about it.

Interestingly, in January 2010, about 7 months before Stuxnet surfaced, I wrote an article stating that if you wanted to conduct a cyber attack against a nation-state in violation of some treaty, then you should target the gas centrifuges with a frequency-based cyber attack to destroy them. That's exactly what Stuxnet did.

As to whether Iran would retaliate against America, it's a complex subject. That's because attributing cyber attacks is very difficult. There are technical ways to trace them. If you know how to do the tracing, you could potentially pinpoint who wrote Stuxnet, from the country down to the organization. There is currently technical evidence that points to the author.

The question is whether Iran is capable of finding the evidence? I don't know. Would Iran launch a cyber attack against the nation-state they *think* did it, even if they cannot prove it? Unlikely. There is a hacker group called the Iranian cyber army that is launching cyber attacks in the world today. They recently defaced the Voice of America's website. They said it was because of some comment Secretary Hilary Clinton made. The real question is, Would that same hacking group launch major cyber attacks against America? I doubt it.

Q: To follow up more broadly, how would you assess the vulnerability of U.S. nuclear power plants—I'm thinking about the so-called nuclear side and the non-nuclear side.

John Bumgarner: There are many cyber vulnerabilities in the electrical industry. Are the nuclear power plants 100 percent secure from a cyber attack?

Unlikely. We have seen some power companies connect very sensitive systems to networks that are not completely severed from the Internet. Some of the networks have been compromised during authorized security tests.

Q: I have two brief questions. The first pertains to the discussion about our neighbors, particularly Mexico. It seems to me that a North American Free Trade Agreement (NAFTA) is something that needs to be explored because American companies were able to enter the Mexican market in banking, insurance, and many other areas precisely because of NAFTA. It seems to be a perfect vehicle to pursue energy security. The second pertains to the role of Russia. Russia has an office right next door to us in Trinidad-Tobago, where every major oil company and gas company has an office. Trinidad has had the distinction of exporting up to 65 percent of our imported liquefied natural gas. Gazprom, from Russia, also has an office there. I haven't heard any references to Russia, though we've talked about China. My question is: What's Russia's role in all of this?

Bernard Cole: Let me briefly touch on the Russian issue. The world's largest reserves of natural gas are probably in Siberia. China is eyeing this with great anticipation. Less than 5 percent of the total Russian population lives in Siberia. In fact, just in the last month, the first shipments went by pipeline from the Russian far east, which is grossly underpopulated. The natural gas fields I mentioned off of Sakhalin Island are being relatively slowly developed because of Russia's nationalistic approach toward foreign companies. These gas fields are potentially a great resource for Japan, but so far that potential is unrealized, thanks to competition with the LNG. The Russia case entails these huge energy reserves accompanied by horrible demographic problems. No one knows how that's going to come out.

Robert Cekuta: Another point about Russia is the fact that the situation will probably become a little more complicated in light of actions they have taken over the past decade in disputes with the Ukraine and Poland.[21] When countries boycott or cut off or slow supplies—and we saw this, actually, with the Chinese on rare earth this past fall—people start getting nervous and look at alternative steps they can take.

You see some of this now in Europe where western European countries and companies are reengineering the gas network to make it reversible or actually multidirectional. Gas doesn't necessarily come in solely from the Mediterranean or Russia or the North Sea, but can be switched around. We've seen some of this right now with Italy, which gets a huge amount of natural gas from Libya for obvious reasons. The Italians wouldn't have been so worried about the war if they could bring gas down from the north. They've actually done work recently on the line coming in from Switzerland. Russian oil and gas supplies are important to the world, and a number of the Russian companies have picked up on this. Lukoil, for example, has become more of a commercial operation.[22]

One thing I would like to add deals with coal. In terms of the energy mix and China, China has just become the world's largest coal importer. The volume is staggering. So again, we have to talk about the energy security notion in terms of all of these different energy streams. There are some positive developments in Russia, and they've got a little bit better sense than they did years ago.

Stephen Kelly: On the question about NAFTA and Mexico, in the NAFTA negotiations, all three countries took certain things off the table, stating they wouldn't talk about them. For Canada, it was the cultural industries; for the United States, it was the free movement of labor and our defense industry; and for the Mexicans, it was energy. While foreign direct investment in many other areas has increased and terrorist numbers were reduced, energy was not one of the areas that Mexico opened up because of the historical baggage.

Q: My question is for Mr. Bumgarner. A lot of the literature in your talk spoke of capabilities. Of course, we know threat means capability plus intent. Can you speak to any intent to disrupt in Estonia and Georgia that you've discovered in any of your research? Have you found any evidence of terrorist groups or nation-states or even criminal organizations, targeting the grid?

John Bumgarner: The report that I wrote on the Georgian and Russian conflict outlined the events surrounding the micro-war. Access to oil in the Caspian Sea was one of the driving forces behind this conflict. From the cyber perspective, whoever launched the attacks did not attack the critical infrastructure. There was some physical kinetic disruption of the infrastructure—a strategic rail bridge was destroyed with explosives. That severed the Georgians from shipping petroleum products via rail from the Caspian Sea to the Black Sea.

The Baku-Tbilisi-Ceyhan (BTC) pipeline which snakes through Georgia, was not damaged inside their borders, but in Turkey an explosion damaged the pipeline. Credit for the attack was claimed by a terrorist organization that operates in Turkey. This was interesting because that terrorist organization had never operated in that region of Turkey before. So no one really knows who disrupted that pipeline system. Russia made sure that it did not target the BTC pipeline at all. They did not bomb it; they did not try to control it. They didn't bomb the seaport used to ship oil. This was a strategic move by Russia.

Concerning cyber terrorism, it is an issue that we've been talking about for years. In 1994 when we held our first cyber war conference in Washington, DC, we talked about launching attacks against critical infrastructures. We also talked about the potential for a terrorist group to launch some type of cyber attack in the future. Someone argued that the al-Qaeda are cave dwellers and really don't have the capabilities. But there are sections of Islamic forums where they talk about critical infrastructure attacks.

If you go to YouTube, you'll see that they do have cyber capabilities—they're using very good propaganda techniques. There's even a good video game on how to plant improvised explosive devices (IEDs). Terrorist groups do have some capabilities.

In the future, we're going to have a problem with a transnational cyber-threat. We're at the point now where traditional insurgencies can form in cyberspace. An insurgency now can function without a real world operational base. It can be a purely digital insurgency or a cyber insurgency.

I've written several articles, one of which is coming out in the *Armed Forces Defense Magazine*, the Pacific Command, about tech-savvy terrorists and how they're going to be changing in Asia and other places, and how they're going to be moving more into cyber capabilities. It's very easy right now for the Islamic fundamentalist groups to become a strong threat to cyberspace. It just takes a little bit of Psychological Operations (PSYOPS) training on their part to move into that realm.

From a nation-state perspective, is there a threat? People consider China a threat to the critical infrastructure. Does China map our infrastructures? Of course. Do we map their infrastructures? Probably. Do they break into our infrastructures to destroy them? No. But we've seen them steal information. When an organization belonging to a state like China breaks into a certain critical infrastructure, they're breaking into an infrastructure to steal information that improves the efficiency of their infrastructure. They're stealing all the features we've worked on to bring our infrastructure up to a certain par. They steal that information, and then when they build their infrastructures online, they can bring it up to the same efficient level that Americans may have taken decades to get to.

Q: There was a lot of press over the years about electromagnetic interference (EMI) threats. Is that no longer an issue?

John Bumgarner: What we are really talking is about electromagnetic pulse (EMP). Some of the current literature on EMP talks about an airburst of a nuclear weapon. There's only a handful of nations that can launch a nuclear weapon. If someone's launched a nuclear weapon against your nation and you're worried about EMP, you're worried about the wrong thing, because you're going to be at war already.

Is it possible to conduct an isolated EMP attack on a critical infrastructure target? Yes. You could construct some type of EMP device in the back of a semitrailer while en route to a location near a critical facility and charge it up and kick it out. Is it easy to do? No. Hopefully, we don't spend a lot of money to get our grids protected from every possible EMP attack. There's been only one known adverse event in the grid that's public, and that was in Canada from a sunstorm. It destroyed a $10 million piece of equipment. This was a freak occurrence.

Some people are worried about EMP events, mostly from sunspots, which have the potential to do a lot of damage. But how do you put Faraday cages around your car, your home, and your television? It's nearly impossible.

Bernard Cole: There's been a lot written about China developing conventional EMI weapons, but the policy and practicality of that are similar to discussions under the last Bush administration when people were talking about conventional warheads on intercontinental ballistic missiles (ICBMs). There's been none developed to date that we know of.

Audience: The phenomenon you're describing is EMP—electromagnetic pulse, which ought to be distinguished from electromagnetic interference, which has a specific frequency. There was another grid incident—a nuclear test atomic bomb that exploded in space above Johnson Atoll took out a string of streetlights in Honolulu due to a big loop of fuse that was pulsed.

ENDNOTES - CHAPTER 5

1. India is also building a port at Sittwe in Burma. See Cowan Thant Zin, "India Confirms Burmese Port Project," *Portworld*, January 9, 2008, available from *www.portworld.com/news/2008/01/70297*.

2. Erica S. Downs, "China's Quest for Overseas Oil," *Far Eastern Economic Review*, Vol. 7, September 2007, p. 54.

3. Bernard D. Cole, *Sea Lanes and Pipelines: Energy Security in Asia*, Westport, CT: Praeger, 2008, p. 81. Also see Robert D. Kaplan, *Monsoon: The Indian Ocean and the Future of American Power*, New York: Random House, 2010, for a discussion of these and many other issues concerning the Indian Ocean.

4. China's National Resources Development Council (NRDC) has established a shipbuilding goal that 60 percent of petroleum imports should be carried in Chinese-flagged hulls by 2050.

5. Interestingly, criminal activity in the form of wildlife trafficking may pose a greater threat than normally considered in criminal analysis: the uncontrolled transfer internationally of disease.

6. "Human Trafficking: The Facts," New York: United Nations Global Initiative to Fight Human Trafficking, available from *unglobalcompact.org/docs/issues_doc/labour/Forced_labour/HUMAN_TRAFFICKING_-_THE_FACTS_-_final.pdf*.

7. "World Drug Report," New York: United Nations Office on Drugs and Crime, 2010.

8. UNCTAD Statistics, Stockholm, Sweden: *SIPRI 2010 Yearbook*, 2010; also see *Defense and Security Intelligence and Analysis Online Database*, IHS Jane's, 2010.

9. See, for instance, Shawn Tandon, "US Ends India Tech Restrictions," Agence France Presse, January 26, 2011, available from *dawn.com/2011/01/26/us-ends-india-tech-restrictions/*. The decision by President Barack Obama was preceded by President George W. Bush's 2008 decision to allow cooperation between U.S. and Indian nuclear power programs.

10. The world's leaders in coal reserves are the United States, Russia, China, Eurasia (including India), and Australia/New Zealand. "International Energy Outlook: Coal, 2010," Washington, DC: U.S. Energy Information Agency, available from *www.eia.doe.gov/oiaf/ieo/coal.html*.

11. Francis turbines, first developed in 1848, are the most common water turbine in use today. They are primarily used for electrical power production.

12. A liquid crystal display (LCD).

13. Heating, Ventilation, and Air Conditioning (HVAC) refers to technology of indoor or automotive environmental comfort.

14. West Texas Intermediate (WTI), also known as Texas light sweet, is a grade of crude oil used as a benchmark in oil pricing.

15. Brent Crude is the biggest of the major classifications of crude oil. It is sourced from the North Sea and used to price two thirds of the worlds internationally traded crude oil supplies.

16. Petróleos Mexicanos (PEMEX) is a Mexican state-owned petroleum company.

17. Felipe de Jesús Calderón Hinojosa assumed office as President of Mexico on December 1, 2006, and was elected for a single 6-year term through 2012.

18. The Cantarell Field or Cantarell Complex is an aging supergiant oil field in Mexico. It was placed on nitrogen injection in 2000. In terms of cumulative production to date, it is by far the largest oil field in Mexico, and one of the largest in the world. But production has declined, and now it is less productive than Mexico's Ku-Maloob-Zaap.

19. Arabian Oil Company (Aramco) effectively started in 1933 when the Saudi government granted a concession to Standard Oil of California (Socal) to explore for oil in Saudi Arabia. In 1936, the Texas Oil Company (Texaco) purchased stakes in the concession, and in 1948, they were joined as investors by Standard Oil of New Jersey (Esso) and Socony Vacuum (later Mobil).

20. Stuxnet is a computer worm discovered in June 2010 that targets Siemens industrial software and equipment running Microsoft Windows. It spies on and subverts industrial systems. Different variants of Stuxnet targeted five Iranian organizations, with the probable target widely suspected to be uranium enrichment infrastructure in Iran. It has been speculated that the United States and Israel were involved.

21. For example, in 2009 a contract dispute between Russia and Ukraine led Russia to shut off gas supplies to seven countries and reduce gas deliveries to several others.

22. Lukoil/LUKoil is Russia's second largest oil company and its second largest producer of oil.

CHAPTER 6

SOLUTIONS

Vikram Rao, William Boettcher, and Douglas Lovelace

This chapter is based on the final panel of the conference. Members of the audience were invited to engage in a discussion over solutions. Douglas Lovelace, William Boettcher, and Vikram Rao served as joint moderators with a view to providing technical, military, and social science perspectives. Vikram Rao led the discussion.

INTRODUCTION

Vikram Rao: This is an experimental session designed to get us thinking more deeply about some of the issues raised over the course of the last couple of days. We'll try to dig deeply into topics that we consider important, or you consider important. Perhaps some solutions will emerge.

We'll invite you to suggest an issue for exploration, we will reframe it, and then moderate a discussion. You will be very active participants in the discussion.

I'm going to start with an issue that has an interesting "deep-dive" possibility. That is Anne Korin's notion of fuel-agnostic cars, which she sees as a solution to our energy problems.

Korin's theory is that it would not be expensive to introduce these flex-fuel cars. General Motors tells us it costs $70.00 to make a new vehicle into a flex-fuel user. Korin's hypothesis is that, if and when we have economic alternative fuels, we'll just put them in our flex-fuel cars, and the cars will run on them. Brazil already has flex cars.

What factors will determine whether flex cars are adopted in the United States? What, for example, about "range anxiety"? Methanol will give you half the miles of some other types of fuel. How will this factor into consumer reaction to fuel-agnostic cars? Fuel anxiety applies to electric vehicles as well. The Leaf will have only a 100-mile range, and that's one of the reasons people may not buy it.

William Boettcher: I would like to add a few comments about the consumer behavior side of it. When it comes to trying to nudge consumer behavior, or military behavior, or behavior of oil companies, we're mainly talking about how to resolve risk and uncertainty or at least communicate to them that the risks are calculated or manageable and that they should take them.

There's a lot of research on how you get people to engage in particular types of behavior. What language you use to present the problem can make a difference in whether particular public policies get support. How you present the problem linguistically can also encourage/discourage particular kinds of behaviors. A lot of this research focuses on whether you should emphasize the gains that can be achieved by this new behavior or focus on the risk of continuing with an existing behavior that's perceived to be harmful.

Depending on the particular behavior and the types of consumers that you're looking at, sometimes it's more important to focus on the gains that might accrue. But more often it's more important to focus on the losses. There's been a lot of research in public health on this. How do you convince people to stop smoking? Or how do you convince women to perform breast self-examinations? We need to think about consumer behavior and marketing as we address these energy security questions. But it is a kind of propaganda. How do you encourage what the state, the military, and the society deem to be appropriate behaviors?

Korin commended low-hanging fruit or fruit that's already in the basket. Some people commended efficiency as the first thing that we should focus on. That is almost exclusively within the realm of consumer behavior. You can get people to be more efficient by offering economic incentives or through good public advocacy, and the appropriate framing of the alternatives that you're presenting to consumers.

Douglas Lovelace: I'm going to provide some military context for this particular question. The first thing we need to realize is that there's a very strong bias within the armed forces of any nation, especially the U.S. Armed Forces, toward effectiveness as opposed to efficiency. The Armed Forces will not tolerate any demands for efficiency that sacrifice one ounce of effectiveness. In fact, the U.S. Armed Forces have proven that they will accept great inefficiencies just to preserve effectiveness or increase effectiveness.

The notion of a tank or a jet or a helicopter operating on alternative fuel is being explored within the U.S. Armed Forces. In fact, the Defense Advanced Research Projects Agency (DARPA) has issued considerable grant money over the past 5 or 6 years to study alternative fuel sources. There have been some fighter jets experiment with alternative fuel sources and maybe some transport airplanes. But the overall bias is going to remain toward effectiveness as opposed to efficiency.

There might be one exception. We can expect the U.S. Armed Forces to remain an expeditionary force for as far as we can see in the future, notwithstanding the 9 years of overseas war that we've been in. Being an expeditionary force means that the Armed Forces have to take their own energy with them, and that's an extraordinary inhibitor to power-projection capability. So to the extent that the Armed Forces are able to identify and develop alternative sources that are easier to deploy, which in the end creates greater expeditionary effectiveness, then those sorts of efficiencies would be welcome.

Vikram Rao: Let me clarify Korin's original position and say she never suggested the agnostic-fuel adaptation for airplanes. She suggested it for civilian consumers, although it could very well apply to military trucks.

Audience: Regarding the comment about whether people will adjust their behavior to accept more frequent fueling, the government should not impose behavioral change or even try to push behavioral change. People are fundamentally economic creatures, and people have different values for their time. For instance, regarding a toll lane, where you can pay a little bit more to get somewhere faster, some people will pay for that and some people will not. In the same way, if you have an open fuel standard where your car can handle gasoline, ethanol, or methanol, some people will say, "Hmm, I want the cheaper fuel and I don't mind refueling more often." Some people will say, "The value of my time is so high, I don't care how much I have to pay for the fuel, I'll pay for the more energy dense fuel and fuel less often." Especially in poor economic times, the bulk of the population falls more to the side of cost sensitivity.

Vikram Rao: Any further comment?

Audience: I question the claim that people are fundamentally economic creatures. They may be economic creatures over the long run, but in the very short term that breaks down. A perfect example of that may well be the past election when analysis shows that many people may have consciously voted against their best economic interests.

Why would anybody bother to go over to alternative fuels unless they are worried about their pocketbook or care about the environment? I don't see any security driver for this. The comment was made earlier that you should change your behavior if you think that carbon dioxide (CO_2) emissions are triggering climate change (which has national security implications). But otherwise, there wasn't a driver on the security side. Going over to alternative fuels may be a personal virtue, as Dick Cheney said, but it's not a matter of national security concern.

The security argument is, in fact, somewhat tied to economics. If we're spending 10 percent of our discretionary income on energy, this has an economic consequence. If the economy is not doing well and if petroleum consumption is the main thing that is fluctuating in people's discretionary income, this increases the need to find alternative fuels. The security argument becomes: What are the available fuels or what is the available infrastructure that can solve this economic problem?

The peak oil question is less debatable than people make it out to be. If it is a finite resource and does deplete, and we therefore go after ethanol or methanol, these are, from a life cycle perspective, fundamentally less economic fuels. If you need a policy that makes them economic to consumers so that they adjust their behavior, you can do so through tax incentives one way or the other. It will change the behavior of consumers, but the question is, how long can you afford to do that before the fundamentals of the different quality in fuels creates an economic problem?

Korin: Regarding the economics, I have a very instructive chart that looks at spot prices of methanol compared to spot prices of gasoline, which are primarily functions of natural gas and oil prices. On a British Thermal Unit (BTU) basis, which means on a per mile basis, for the most part over the last 3 years, methanol is significantly more economic, meaning less expensive, per mile than gasoline. With ethanol, it's a different story.

Also, in regard to people being economic creatures, how many of you have driven around the block to get gasoline that's 5 cents cheaper? In general, a lot of people do that. You cannot argue from election results that people are voting against their economic interests because people have different perceptions of what the results of an election would provide for them.

Audience: I'm from the city of Raleigh, North Carolina, and the city has 5,000 employees, a third of whom are concerned about the rising gas prices and not being able to come to work.

Vikram Rao: Let me make one point on the military aspect. Let's say that 10 percent of the civilian vehicular fleet switches to something other than oil-based fuel. In that case, there's more available for jet fuel. So from a military security standpoint, yes, that is advantageous. The conventional fuel will be more available for the military if the public uses unconventional fuel. I see that as a weak but non-zero connection.

Audience: Let's assume that this all works and that the methanol drives down the price of gasoline and petroleum. What happens to those nations that are living on profits from oil exports? Does that make our world more unstable for our neighbors?

I think this is a fascinating topic. Stripping oil of its strategic status is a necessary but not sufficient condition for a lot of different things. If you look at parts of the world that are very poor or hoping for an upward trajectory, you focus on the liquid fuel side, opening vehicles to fuel competition, and removing the trade barriers and tariffs. You're enabling the poorest people of the world to sell into one of the most lucrative markets, liquid fuel, by producing alcohol fuels. That's a very good thing. It also can develop a healthy economic interdependence—for instance, if we remove ethanol tariffs for larger parts of Latin America. That's on the positive side.

If you look at what happens to countries that are primarily oil exporters and ask yourself, is this going to destabilize them? I believe that just as we are addicted to oil, they are addicted to petrodollars. Would you rather be a Saudi prince than a Jordanian prince? Probably Saudi. But would you rather be a Saudi subject or a Jordanian subject? Probably Jordanian. The difference is, when a country lacks resources and can't stick a straw in the ground and extract an enormous amount of wealth, then it does have to enable its entire population to be able to be educated, not just the men; it does have to enable more of a participatory type of governance—maybe not democracy as we think of it, but something more participatory. If it doesn't have to do that, then you get Saudi Arabia.

I don't think the world's lessening of biofuel dependency will be an easy transition for the Middle East. I certainly don't think it'll solve all the Middle East's problems, but I do not think we will ever get to anything resembling normality in the Middle East until they are weaned from petrodollars.

Eugene Gholz: It's an interesting question if you're an academic gazing at world politics, or if you're an altruist interested in development and bettering people around the world. But from a policy perspective, it's pretty remarkable to be thinking about the poor Saudis who are going to be really hurt if we do something to reduce the price of oil. From a policy perspective, we don't even think that hard about how to keep up the prices of American exports or how to reduce the prices of American imports. To try to have an income-preserving policy for the people of another country, such as Saudi Arabia, when we don't have one for the people of our country (and I don't think we should have one for the people of our country) strikes me as a tough sell in the policy community.

Audience: I've just had some discussions with the Saudis on some of these things in their country. A couple of things that are going on there are quite interesting. One is, when you talk to people in Saudi Arabia, including in the government, you'll find they are looking to develop a more rational use of energy within the kingdom.

Right now energy prices are subsidized. The Saudis are asking themselves how to wean themselves from this. They're looking at this in terms of the rapidly rising demand for crude oil within the Kingdom where it's burned to make electricity. Why is electricity needed in the kingdom? For water and air conditioning.

What is critical for us in, say, Maine, is keeping warm in the winter. But in Iraq and in other Middle Eastern countries there's real political pressure to get the energy that's needed so that they can have air conditioning to make life bearable at 110, 114 degrees in

Basra or Baghdad during the summer. So the Kingdom, Iraq, and the others are looking at ways to try to limit their consumption as part of their development.

In terms of Saudi Arabia, we are seeing some remarkable changes. I drove by a brand new university for 30,000 women being built outside Riyadh. There is a realization of the need to change, and one of the things that the King is trying to change is to focus on the tremendous resource which is not being utilized. Women constitute 50 percent of its population.

A related issue in Saudi Arabia, in Jordan, in Yemen, as well as throughout the Middle East, is the level of chronic unemployment and underemployment. What we've seen in recent weeks in Egypt provides us with a perfectly good example, indeed a brilliant example. In Egypt, 30 percent of the population have a university degree but no job. There is an urgent need to create jobs, to release their entrepreneurial spirit. The oil curse is a factor. But some of these factors are factors across the whole of the region whether they have hydrocarbons under the ground or not.

Audience: The Saudis are doing quite well, getting $40 to $50 a barrel. But let's go back to Douglas Lovelace's point and talk about the military's effectiveness and efficiency. Under the new dynamic of facing a $100 billion reduction in the federal budget for the Department of Defense (DoD), certainly we're going to look at some efficiencies, at least some basic consolidations, and some other ideas on the way we make the military more effective.

Douglas Lovelace: So far the thrust of the DoD has been toward increasing the efficiency of management processes, acquiring material systems more efficiently, running posts, camps, and stations more efficiently. That wasn't really the effectiveness versus efficiency that I was speaking of. I was speaking of it in a warfighting context. For example, if a U.S. jet fighter has to sacrifice 20 knots to burn a flex fuel and that 20 knots make it not quite as good as the new Chinese jet fighter using conventional jet fuel, the U.S. Air Force would justifiably resist that.

That said, however, if we look at the National Security documents that have been produced by this administration, they all describe a particular vector that the U.S. Government is taking with respect to providing for national security. We can start with President Obama's National Security Strategy, go on to the Chairman and Joint Chiefs of Staff who just recently published the national military strategy, and to the Department of State, which finally published its *Quadrennial Defense and Diplomacy Review* (QDDR). This vector is to deemphasize large, costly, energy-burning land forces that would be projected around the world and to place greater emphasis on naval and air forces, i.e., over-the-horizon forces, the deterrent value being a fleet out there "somewhere," and the enemy not knowing exactly where it is at any time.

In addition to that, a few years ago we were talking about military power, and then a couple of years after that we started talking about soft power to complement the military power, and we all know what that is. Then we replaced both of those terms with smart power, as if military power and soft power were dumb power before. Now in the new QDDR from the State Department, the Secretary talks about civilian power in juxtaposition with military power. I think it was Secretary Albright who said U.S. foreign policy is based on the application of military power backed up by the threat of diplomacy. Thus if this vector continues and we move away from the military as the principal instrument of national power, you're going to see the U.S. Armed Forces reduce their reliance on energy.

Vikram Rao: On a somewhat different note, I wonder if the military, because it's focused on pragmatic, performance-based goals, would be a more rational consumer of different forms of energy. If you could demonstrate a performance standard, I think the military would accept it. I don't think that the general population is that rational. I think they would have much more emotional attachment to their vehicles. You'd have to overcome all sorts of other barriers to acceptance among civilians. You'd have an advantage dealing with the military.

Audience: Let me address very briefly the question of a nuclear navy. The Navy has proven that nuclear power can be operated safely, but it's at a very high cost in personnel training, safeguards, and so forth. These costs are probably too high to be widely applied to the civilian sector. The second point is that the Navy has decommissioned all its nuclear-powered surface ships except for the aircraft carriers. That was because of the high personnel training costs. Since the Cold War, it costs too much in terms of personnel to maintain nuclear-powered surface ships other than aircraft carriers.

Audience: When you talk about energy supplies in the context of energy security, there are many factors to consider, and sometimes these factors seem to be in conflict with each other. For example, civilian security seems to be very different from national security; some civilians may be concerned about cost efficiency, and you may have different concerns at a national level. You need to have a balance between policy, which is more or less civilian-geared, and strategy, which is more or less national security-geared. How do you balance the difference between policy and strategy? In Turkey, for example, four out of 10 vehicles are running on propane. The Turks depended completely on imported oil supplies and wanted to shift away from it. Somehow they decided they were going to go with propane gas. We're looking for an alternative fuel here in America, but how do you balance the gap between policy and strategy in the energy security field?

There are different ways of looking at energy security. As a civilian, I may be more concerned about how much I pay at the pump than other things. I really may not care about what would be the impact on our allies if we shift away from importing oil. But at a national security level, you do ask these questions.

Audience: There have actually been several issues raised earlier that I'd like to pursue. First, let me return to the discussion about the military. If you think about the military, not every Soldier or Sailor needs a cruise missile. The military makes decisions based on its needs. So maybe 20 percent of the military needs high-BTU fuels, but maybe 80 percent don't. So it seems that the military could also develop a very rational policy for deploying fuels in different arenas and different uses. For front line deployment, you might want the really high BTU fuels, but on the back lines, riding around a base in the United States where you probably don't go 100 miles a day, you could probably use methanol pretty easily. That's one of the issues.

But I'm especially interested in the broader issue about how to change the use of energy resources in the United States. Is the market in the United States big enough to have an impact on the cartel, and what percentage of the market do you have to change in order to have an impact on the cartel?

That in itself assumes that our goal is to break the cartel. So the further question I would have is: What's the implication of that for the global economy as opposed to global energy use? And, if you changed the dynamics of energy policy around the world and energy costs and markets, wouldn't this potentially have a broader impact on the global economy? Then, speaking of security, wouldn't we be more worried about destabilizing China, Russia, India, and Pakistan than destabilizing Kuwait?

Audience: Commentators have spoken about the numbers. Nobody is talking about breaking cartels because nobody believes that alternative fuels will give you more than 20 percent penetration in 20 years.

A fuel choice would not break the cartel soon. It would take time. Remember, fuel choice is not the same as mandating a level of fuel. It's opening the cars to different kinds of fuel so the market can decide among different energy commodities. Even if you're not talking about displacing 80 percent of petroleum fuel, customer choice will essentially cap the price of oil at that point where liquid fuels are profitable, so it'll cap it at about $55.00 a barrel. When you have electric vehicles going into the market in numbers, it'll change it further. So it'll have a very significant impact even before you get to a very large penetration of alternative fuel. The key is really for the cars to be able to handle the fuel. Then you send a signal to investors that they can drastically expand capacity.

When our speakers have talked about the limited number of alternatives we can expect to see, they were focusing on renewable fuels. But if you look at how China has ramped up coal to methanol, it's unbelievable. If you look at the potential of converting natural gas to methanol, it's also very large. When you look at the introduction of these alternative energy sources into the transportation sector, you're encountering something very different from renewables.

Vikram Rao: Thus even if we don't change the balance in this country by a huge percentage through alternative fuels, they do change market choice and therefore the mindset of the producing nations, and they have an effect on the cartel.

Audience: The point has been made that the markets will work. If alternative fuels become widely available, their prices will bring about an equilibrium with the petroleum fuel prices which will be somewhere above the cost of extracting oil from the ground in the quantities that equilibrium hits. It'll be somewhere above what it costs to produce those alternative fuels.

The price of ethanol in Brazil is the same as the price of gasoline on a per mile basis. The markets are working even in Brazil, and that would happen here as well. The larger the alternative fuel sector gets, the more influential alternatives will be on that equilibrium price.

I would ask, however, about multi-fuel vehicles: Will we have the necessary infrastructure in place to make them work? If you can't find the alternative fuel, it doesn't matter whether you have a multi-fuel vehicle or not. That infrastructure will have to be used pretty close to capacity in order for it to be maintained or expanded. We'll thus be forced to make some decisions about what kind of alternative fuels to introduce to the market. I like the idea of fleets and small groups of vehicles being converted. That's likely the transition path we'll follow.

It should be the responsibility of every state and local government to drive this change.

Douglas Lovelace: One clarification, responding to the gentleman's comment that the DoD or U.S. Armed Forces could achieve efficiencies from the use of alternative fuels in some noncombat type vehicles. Just recently in Pennsylvania, I drove a car out of our General Services Administration (GSA) fleet, and it had a big placard on it reading, "Takes E85." But in the whole area, there was only one station that had an E85 pump, and it wasn't convenient for me to use it. But such an experience certainly draws attention to the point made about the infrastructure.

About a year ago, I rented a car in Sioux Falls, South Dakota, and the rental car agency made me sign a statement promising not to put E85 in the car. If I had put it in, there would have been a big penalty — $500.00 or thereabouts. We have a ways to go with infrastructure.

Audience: The speaker said that in the private sector change needs to be demand-driven. That means we have to have the flex-fuel vehicles out there, which means we force the production of them or force the consumption of them or force the purchase of them. Do we have the capacity to have a massive cash-for-clunkers program which defines what can be purchased to replace those vehicles that we're taking off the roads? We'd need something like that to get the kind of massive transformation of the consuming fleet we are speaking of. On the supply side, the only means we have, other than mandates, to effect change is to have a tax credit system that could be used to encourage installation of conversion kits.

Vikram Rao: An earlier point was that regulations/legislation should be put in place to make sure that every new (not retrofitted) car has this feature. This would provide potential capability and encourage the production of alternative fuels. It wouldn't be a question of incentives, but of legislation. If you believe GM's number, $70, then it would not appear to be a high price to pay.

Audience: You're asking people to be proactive and to impose costs on particular constituencies, and that's un-American.

Vikram Rao: It happens all the time.

Audience: There's no crisis right now, but we need to adapt in this way, and it's going to be relatively costly. If you're going to tax people or you're going to have tax credits, you're going to be moving behavior, and you've got to have a huge public education campaign that's very carefully crafted, that convinces Americans that this is the right path to travel on. Otherwise, you're talking about avoiding long-term risks and experiencing short-term pain, and that is not what we're good at. This is why climate change is a very difficult issue for us.

Audience: There's a big issue here. People in this country are very much economic animals. They want to minimize their costs so they have more disposable income to do what they want. People will spend a lot of money, disposable income, on social status and so forth, but they will not do that buying a big SUV if the cost of that SUV starts wrecking their budget. A good point was made about how behavior started to change when the amount of money spent on transportation by people went from 3 to 6 percent to 8 percent, and started getting up to 10 percent. Then they say, "That's beyond my budget and I will give up my SUV, or I might buy a more fuel-efficient car."

We've all probably sat down and done the calculation. "I can go out and buy myself a plug-in Prius and save a bunch of money." But the plug-in Prius is going to be $28,000 when it comes out. How much am I going to save at $3.00 a gallon? I'm going to save $10,000 over 5 years? That doesn't pay for a new Prius when I can buy a traditional car for $18,000. People make such calculations. They may not always be made with perfect accuracy, but people think in this way. Unless the cost gets high enough to be painful and to cause them to devote much more capital in cutting their longer-term fuel costs or their short-term pain, then it's not going to happen.

The point was made very nicely that the cartel understands that. If fuel gets too expensive in the United States, then people start doing something about it. So the cartel has been fairly good at keeping the price below the pain threshold for the United States and the rest of the world. The only way around this is for the U.S. Government to do something that causes the perceived price of fuels to go up.

Our automobile industry has a long history of being regulated and subject to mandates. We've been required to put on seatbelts, crash-resistant bumpers, and deal with emission standards. Implementing a change like this should be trivial.

Anne Korin: Let me just for a moment enter the political arena, because the political arena really reflects what's realistic from a public policy perspective. In the last session of Congress, the cost of an open fuel standard (in terms of automaker implementation per vehicle) was modest (under $100.00). As a result, both conservative Republicans and Democrats supported an open fuel standard bill. You'll see the same thing in this session when it's reintroduced. It's something that's politically feasible.

I also want to address the infrastructure issue, which is very important. A car is on the road for 16.8

years—that's the turnover time of the fleet. If you put an open fuel standard in place, 2 or 3 years later you will hit a point where, let's say, 15 percent of the vehicle fleet—not new cars but the entire fleet—is flex fuel. It is only at that point that a fuel station owner who has 10 pumps can make a business case to either retrofit one of those pumps to dispense alcohol or put in a new one.

There's a $50,000 tax credit today, which more than covers the cost of retrofit and covers 80 percent of the cost of putting in a new pump, but nobody's going to take advantage of it until there's a significant number of cars out there that can use the fuel. Once the fueling infrastructure is there, oil prices will become volatile. At that point, if the oil price goes up, you will want to be able to use this other fuel. If you have a used car that can't use that fuel, then you will go to entrepreneurs who own used car lots to do conversions and offer used cars that are open fuel standard compliant. You can also envision a business opportunity for, say, Midases or Meinekes to do $300 or $400 dollar conversions.

If you look at the cost savings for an average car that is driven 12,000 miles a year on average, with the average fuel efficiency that we have for today's cars, as long as the spot price of methanol is around a $1.00 to $1.20, and gasoline is around $3.50 net, you'll end up saving $300 or $400 a year. Within less than a year, you pay back that additional very small marginal cost of the fuel flexible car.

Audience: I'm still struggling to see the security benefits coming from all of this. In fact, I didn't hear any upside for the security side of this push for alternative fuels. I only heard a down side which was the doubling of food prices around the globe and resulting riots in countries. That is a security issue, and we should be concerned about that. But otherwise, it might be virtuous of me to buy a Prius and I should feel good about that, but it doesn't have any positive implications for national security.

I have a question. Is there any data showing what percentage of people we might expect to buy a flex-fuel car before it becomes economically indicated? That is, before there are obvious incentives? Who will be the first 15 percent that buy these cars in the first 2 years?

It's like a seatbelt law or an airbag law, it's a very inexpensive feature. After you've made your car flex-fuel, it looks exactly the same. It just has a fuel tank that can handle a variety of fuels. You can fuel it with gasoline if you want to. If you have a seatbelt in your car, if you don't want to use the seatbelt, you don't use it.

The security benefit comes from economics. We did a rough calculation on the cost of importing oil to run our cars and all the externalities that are associated with it. It's about a trillion dollars a year. If you can develop enough alternate energy sources within the United States of all types, you're talking about a cash flow savings about the equivalent of the deficit just from adopting this kind of technology. That's the big issue. We're not the largest oil producing country in the world, and if we come up with energy alternatives, it hugely protects the cash flow of our nation.

Anne Korin: I just wanted to respond to the question about the early adopters of flex-vehicles. We already have flex-fuel vehicles that can operate on E85 or gasoline. They're provided to the motoring public at no additional cost—we already have seven million on the road today, including about 180,000 thousand in North Carolina. We have only 14 E85 stations, however. It is a big challenge because there is not the economics in place for the petroleum marketer. We have to have grants to offset that cost. We did have a tax credit of $30,000 that was dropped at the last session.

But to get to the point about how many early adopters we might expect, I think we have an example with hybrid vehicles. They've been on the market for over a decade now, and recently about 3.4 percent of new car sales were hybrids. But they cost a lot more than your average car. They don't face the infrastructure challenges, but they do cost a lot. The price premium is $5,000 or $6,000 in some cases.

Audience: I have a fleet of five compressed natural gas (CNG) vehicles here in North Carolina. Chapel Hill just initiated a program of grants to improve the energy efficiency on one's house. I just applied for the funds to install a piece of equipment that goes inside the garage that you can use to refill your CNG vehicle in the driveway. Also included is an operations and maintenance manual for the CNG pump itself, a brief on the advantages of using CNG fuel, and a portion of the city code of the City of Chico, California, which mandates that all new garages have a provision for a natural gas vehicle refueling appliance. I'm taking this step just to see what the environment is for actual implementation of CNG.

I have a very basic question from a social science perspective. What is really the nexus between energy and security? The reason I ask that is, after the 1973 oil embargo, we had cars downsizing or getting smaller. Then after a few years we forgot all about it, and everybody went for the gas-guzzlers. Recently we're talking about that again, and there is a lot of talk in the literature and in the news about having to move away from Middle East oil dependency. It seems to me that if we can import it cheaply, then that's the way to go.

So why are we trying to go from Middle East oil to, let's say, Canadian oil resources where it's not that easy or cheap? Why is dependency a concern in the first place? Why are we focusing on having to get out of the Middle East? I understand it's a volatile place, but we have as friends in the Middle East the dictators, not the democracies, and dictators are much easier to control. We also learned after 1973 that, just as we want to get the oil from them, they need our market to sell the oil. We also learned that one easy way of dealing with the Middle East oil cartel was to focus on the dollar, because they invested everything here.

Why, if we leave the Arab/Israeli conflict out of the equation, are we so concerned about our dependence on the Middle East? While we are busy making it clear we want to become less dependent on them, they are not sitting back. They are finding alternative markets in China and other places.

I think some of the answer to that can be found in the QDDR that Secretary Clinton issued in December 2011. The fact is that when we're buying foreign oil—and I don't care whether it's from Saudi Arabia or Canada—it means that money is leaving the United States. One of the problems we've had is a very large, almost chronic, balance of trade deficit in this country. We complain about losing jobs because of manufacturing going overseas, but we're buying stuff from overseas every time we go to the pump. Let's say about 60 percent of the energy that we're using in terms of oil is being imported. That's money we are paying out to somebody else.

The other thing to remember is that oil has been cut off in the past. We're in much better shape today than 1979. We've got reserves. We are not in the situation where we face gas lines. But the price of Brent oil, which is a good bellwether crude, is going up. Why? Because of concerns about Libya, because of concerns about the Middle East.

Audience: We should not put ourselves in a situation where we are subject to the vagaries of decisions and actions taking elsewhere in the world. This is an issue we need to deal with. That's why the President, the Secretary of State, the Secretary of Defense, the generals, and the diplomats are all saying the same thing—that we need to be doing what we can to limit our dependence on this foreign input into our economy. Try running an economy without energy, you can't do it. We need to find ways in which we can meet our fuel needs at home—gas and other things are already helping.

I'll second that. U.S. security is tied to the U.S. position as an economic power in the world. It may be, however, that the alternatives, even the fossil alternatives (say, the Canadian oil sands) can't scale up and match the price of Saudi Arabian crude. I agree with the previous speaker that we should make cars able to run on different kinds of fuels. This is a part of the solution. Electricity, alternative fossils, or renewables give consumers choice.

Audience: A previous speaker asked, Could a U.S. city like Raleigh, North Carolina, survive for 6 months without a functional electrical power grid? His answer: No. I'd like to qualify his answer.

It's certainly true that life, as we have come to live it here in Raleigh, would not be viable for the 6 months ahead of us without the grid. However, most of us would indeed survive the experience and might, in fact, come out as more rounded people. I recommend a movie you might watch on this topic called *Zero Impact Man*, a low-budget documentary film made by Colin Beavan about how his family lived in a ninth floor apartment in Manhattan without the circuit breaker connecting him to the grid being on. They switched themselves off for 6 months. It was not without inconvenience. In terms of George H. Bush's comment that the American lifestyle is not negotiable, if you insist that peak oil is not real, you can continue to refuse to negotiate. If you want to negotiate, then there is no one to negotiate with because it's actually the global oil petroleum production rate, not the reserves, that matters. It's all about the rate, it's not coming out of the ground as fast as it used to. Recall the propane now being used in Turkey. Then there was the taxi driver a colleague of mine encountered in Serbia—he had a propane tank, a natural gas tank, and a regular gasoline tank in the same vehicle; he had adjusted. He changed his lifestyle. When they have to adjust, Americans will change their lifestyle. They won't negotiate, they'll just be forced. So the question is, Will that happen? It's a longer-term question.

Audience: This is a related question. It seems that the oil industry monopolizes the whole line—exploration, refinement, distribution, selling, and everything. That has implications for this discussion. First, I don't understand why the industry has this monopoly, and, second, what are the implications of this monopoly?

First, you really have to distinguish between the International Oil Companies (IOCs) and the National Oil Companies (NOCs). Most of the world's oil is owned by NOCs, essentially by governments. The IOCs like Chevron, Texaco, Exxon, or Shell, own only 6 percent of world oil reserves and have access to under a quarter. If you think about where they make their money, the NOCs are making their money upstream mostly. But the IOCs are more and more like the Red Queen in Alice in Wonderland—they have to run just to stay in place. They're not necessarily allowed to explore and lift oil in the places where it would be cheapest to do so overseas. That's why a lot of them are becoming more and more natural gas companies. They make their money along the entire supply chain, as well as the processing part and the distribution. For them, producing fuel from natural gas or distributing fuel from coal is no less appealing. They're riding a very lucrative horse, and they're not going to get off it a moment before they have to. Thus I don't see the IOCs at all as an obstacle; the NOCs, that's a different story.

Audience: We need to understand that alternative fuel is not the same as renewable fuel, just like a cow is not the same as livestock. Just as a cow is a subset of livestock, renewable fuel is a subset of alternative fuel. So when you're thinking about alcohol as an alternative fuel, you're thinking about fuel from coal and fuel from natural gas. Nobody eats coal or natural gas. You have to think about the whole spectrum. Very frequently those fuels are much more economic on a per mile basis than either petroleum or fuels from biomass. Now there are many other factors that come into play for people concerned about environmental issues, but if the concern is economics, one simply cannot ignore the broadest spectrum of fuels.

A speaker earlier said that we're spending a trillion dollars a year on these fuels. I have a solution. If we went over to diesel, where you can double your mileage per gallon, we could be there tomorrow if somebody would legislate it. If what's driving us is the notion that we should spend less money on importing fuels, then you can solve that with a stroke of a pen with a fuel standard.

Big oil seems extraordinarily efficient, delivering a fantastic product right to my doorstep, a gallon of gasoline. It seems like it actually should cost 10 times more. The amount of energy you have in a gallon of gasoline is extraordinary. The fact that it can be safely distributed around the planet makes it very tough to displace. It is the perfect fuel.

The only objection you can have is that it loads up the atmosphere with carbon dioxide. I'm actually not as concerned about that as the earlier speaker who made such a passionate case for the dangers of climate change and global warming. Clearly having more CO_2 does lead to warming, but the implications are much less clear for the climate in the long term. But big oil is doing a pretty good job today, and we should not disturb them too much.

Vikram Rao: Small clarification on fact. Diesel will get you 35 percent more mileage than gasoline (not 100 percent), and 10 percent of that is because it's got more calories; and 20 percent of that is because of the high compression engine. If you used a high compression engine, other things can happens.

Audience: A comment on the U.S. lifestyle issue. Europe has a great quality of life but has a lower carbon footprint. That is because petroleum is more expensive there, and public transit and smaller automobiles are therefore more attractive. We need to think about those types of things in the United States. The reason why it's difficult is that we have a very weak government system here. It's also because very powerful corporate interests, including big oil, prevent us from making needed changes.

John Bumgarner: Let me return to the claim made earlier that we could survive a major attack on the grid—let me challenge that. Some individuals could survive, but if we are talking about a long-term event, it would be very difficult. We had a major snowstorm, with the power where I lived going out for about 3 weeks. Most people packed up and moved out of the area and went to hotels and relatives. My wife and I bunkered into our home because I have a lot of field equipment, things like sleeping bags, whisper stoves to cook on, and chemical lights. I built an igloo outside on the deck to keep our meat in. But if it had been a summer event, it would have been more difficult: it would have been too hot to deal with.

In Estonia fairly recently, the government ran their first national security exercise, and I was one of the authors of that exercise. If they had some type of event, say a major power outage, how could they deal with it? The reason they asked this question is that Estonia is the most wired nation in the world. They're really serious about Internet technologies. They use only debit cards or credit cards, very little cash, and no checks. So when they have an event over there that knocks out the electrical power, it's difficult to buy groceries or do any type of commercial transaction. It would mean no online banking and no communications. At the time, we tried to figure out how they might develop a national emergency program to issue checks to the entire population of Estonia so they could write checks on their bank accounts.

This is a major problem for some areas of the world, and we really need to think about it. Could we hunker down for months at a time? If the State of California, for example, was the target of a successful cyber attack, it could impact the whole U.S. economy.

Audience: Back to Colin Beavan's documentary family. It was inconvenient without heat in their apartment during the winter, lots of sweaters, lots of cuddling together, lots of telling stories and singing, but they did pull through. Colin used a solar panel on the roof to continue to blog throughout the experience.

Here in Raleigh, we might experience our forced return to 19th-century technology as a refreshing opportunity to reconnect with our neighbors, ourselves, and perhaps with nature. We might realize, therefore, that our connection to nature is actually more essential to our continued viability as a species than is our participation in the conveniences that are offered by our modern industrial civilization.

Vikram Rao: We're headed the way of Estonia. We're going to be talking about a national health policy in the United States where we'll have personalized health and all the electronic involvement that goes with it; this specter is being raised at a time when our dependence on electricity (and energy) is going to become an issue. Solution?

Audience: Resiliency. If it comes to resiliency, we should make maximum efforts at conservation in all forms and fashion.

Vikram Rao: There was some discussion about energy efficiency. Arnold Schwarzenegger's mention of it in a recent conference was alluded to. California tried a grand experiment in 1971, enacting several electricity conservation measures. California had a per capita consumption of electricity of 6,000 kilowatt hours per capita annually. The rest of the country had 6,600. So they were using less. By 2001, California was up to 7,000 but the rest of the country was up to 12,200. That is a 40 percent difference. One of California's measures was to mandate that standby power, electronics, and TVs had to consume less than one watt. It used to be 6, 10, 15 watts. That legislation was costless. That particular change in the end—not in the beginning—but in the end had zero additional cost. I suggest that energy efficiency does not necessarily equate to consumer privation.

Audience: I want to make a plug for a series of three workshops we're holding in North Carolina regarding fuel economy for your vehicle. Just by changing your driving habits, you can save up to 30 percent on fuel. This is a practical application. The workshops are free. The first one's going to be right here in Raleigh on April 13, 2013.

Audience: I want to comment on the efficiency issue and bring the conversation back to the political discussion we began earlier. In Congress, the House Republicans recently ended Nancy Pelosi's Green Initiative, which installed biodegradable plates and utensils in the House cafeterias. When you have that kind of political climate, you've got a problem. We can sit here and declare that if everybody would change their light bulbs, we could reduce energy consumption by 20 percent. The political reality is such that improvements have stalled. We have to figure out a way to sell common sense, which is frustrating. We're in a political climate where we're ignoring good ideas.

Audience: People talk about the compact fluorescent (CF) light bulbs.[1] But these CF light bulbs are a problem. My electrical company in its great wisdom sent me over a year ago three boxes of CFs. Those three boxes of CFs are still sitting on my floor. I'm waiting for light emitting diode (LED) technology to catch up. Once LEDs come into fashion, I will deploy LEDs. But CFs I will not deploy in my household.

I don't think that we've taken into consideration their impact on the environment They might save us energy. Australia, is going to go to all-CF soon just like America. They might save us energy but no one's really thought about what happens at the end. When the Environmental Protection Agency (EPA) tells me I have to open my window to air out my house after I break a CF, there's a problem. If I drop an LED, the EPA doesn't tell me that I need to air out my home. I don't think we've done the environmental impact analysis. If you replace all light bulbs in America, you just can't divine the environmental impact of what this is going to be for America. It may save us some money, but the environmental costs haven't been taken into consideration.

Vikram Rao: Yes, the CF bulbs have nasty stuff in them, but does the EPA actually suggest that if you break one, you've got to open windows? What if you don't have windows?

Audience: CF bulbs have a percentage of mercury in them. We heard earlier about effectiveness versus efficiency and it applies not just to the military but to our individual lives. The law that outlawed incandescent bulbs in favor of fluorescent bulbs is an example of a failure to recognize that. People don't equate light from incandescent bulbs with light from compact fluorescents, they're two different forms of light, and they're not equally effective. If you look like a ghost in the mirror some morning when you have a fluorescent bulb, it's because it's not the same as an incandescent bulb. Consumers judge on quality as well as price, and effectiveness as well as efficiency. Some of the policy problems we've had are where people have imposed measures that failed to recognize this.

If there's zero cost and some marginal improvement in effectiveness in mandated measures, then nobody's going to object. But if there is an imposed loss of effectiveness or quality in some way, then there will be objections. We have lots of room for enlightened policy, and we can also build on incentives.

The advice to open the window is perhaps a bit of overkill. It's not a bad thing to do. But all of these fluorescents up here contain mercury too. So they're no different than a compact fluorescent. There is a range of issues that come into fluorescents. In California, they have all fluorescents there. You turn them on, they come on really fast, they're bright almost immediately, and their color is good. But if you go out here and you buy a compact fluorescent, the color might be good, it might come on fast, and it might take a long time to warm up to full brightness. It's the technology and the market catching up with the regulation.

Regarding compact fluorescents, you can buy at Home Depot color balance. You can get all kinds of different colors, depending upon the phosphorus they put on the inside of the tube. I started using them about 12 years ago living in Boston when the utility offered them for free. They didn't send them to me but made them available for free, and I've used a lot of them since then. I've put them in recycling so that mercury can be recovered. I've broken one or two, but I'm not too concerned about the small amount of mercury exposure I've gotten.

The EPA says right here—I'm reading this—that if you break a fluorescent CF, it's recommended that you open the windows and step outside the building. You should also remove all pets from the room for 15 minutes, and you should turn off your central air conditioning or heating systems because you don't want the mercury vapors from the light bulbs going through your home.

Audience: At least you don't have to call HazMat. The environmental impact issue makes me think about the environmental impact of using lower BTU fuels. Aren't you, in fact, producing more CO_2 because you're burning more to get the same bang for your buck? Isn't there an environmental impact of using methanol and ethanol?

Actually not. The reason this kind of fuel has a lower BTU is that there is less carbon in it, and therefore the amount of CO_2 is less.

Audience: To get back to the larger question about energy and security, can we achieve energy security without any adverse impact on the rest of the world?

As earlier emphasized, the United States, and in fact much of the world, has a pretty significant degree of energy security — as if you could actually measure higher or lower energy security! But our energy security situation is better than you think. People are alarmist to a degree not supported by underlying evidence. It depends partly on how you choose to define energy security Energy security is essentially about protection against sudden increases in price, or sudden disruptions of supply. In other words, it's about short-term acute changes.

There might be a long-term trend to higher energy prices, whether due to peak oil or, much more likely, increasing demand in the developing world, notably China and India. There is more competition, more demand. Meantime, supply is increasing at some other rate. So prices may go up. That may be a concern. There may be environmental concerns which also lead to higher prices.

But the security issue doesn't seem to me to be a big one. Of course, there are other energy security issues besides the price spikes or supply disruptions. But we are largely insulated against these already. One can argue about whether there is a security risk in that some amount of energy money goes to governments we don't like.

But the basic thing to start with is that America's actually a wonderful place. What great news. We're very secure, we're very rich, and we have a society capable of dealing with lots of problems that confront us. Whether one's concern is over energy security, or whether your neighbor is going to invade and enslave you, or anything else you can think of, I'd rather be here than anywhere else I can think of. And if you think you'd rather be somewhere else, we don't actually require you to stay. What a great society. I think we are energy secure in a significant fashion.

Vikram Rao: The International Energy Agency (IEA), if I remember correctly, defines energy security as having energy that's affordable and respects the environment.[2]

Audience: Reliable and cheap.

Vikram Rao: So that's what they define as energy security. That's not the military definition, but there's a military impact as well. We've had a lot of discussion but perhaps not focused enough on the military dimension. Comment?

Audience: In 2005 I took part in a study that Andy Marshall did in an analysis office of the Pentagon, unclassified, and his question was very carefully, narrowly phrased. He asked: Could we interfere with the sea-borne delivery of petroleum products to China to a sufficient degree to affect a decisionmaking process in Beijing. I thought that was a great way to phrase the question. We didn't ask whether we could do anything like block the transports. My answer to that careful question was, No. It goes to the larger point of the internationalization of the energy industry. The industry as a whole has become too internationalized to be the military point of leverage it was, say, during World War I, or particularly World War II in the North Atlantic. I think we're beyond that right now.

Vikram Rao: Comment?

Douglas Lovelace: Two points. One that has been alluded to by a lot of the participants today, i.e., that the distribution of energy ownership certainly has geopolitical ramifications. A good example of that took place when the Soviet Union disintegrated. At that time, the Russian Federation went into an economic tailspin for a while, and their foreign policy became very passive. They engaged in almost no foreign policy initiatives that were inimical to the United States. In fact, for a period of time their interests seemed to be very consistent with those of the United States.

Then they got their petroleum and gas legs underneath them, the price of oil and gas went up, and their foreign policy became very activist again. The possession of oil reserves and gas reserves in other countries of the world forces us into situations where we must accommodate or at least coexist with leaders that we otherwise probably wouldn't choose to. So there are some geopolitical ramifications to the control of energy.

Some of the speakers have talked about the security aspects of increasing competition for energy sources which is likely to take place given the economic growth of India and China. I personally don't think that has a military dimension. The United States needs to pay close attention to and manage this challenge. There will probably be very difficult times in the future as China and India emerge — they are major claimants for energy resources. However, the other instruments of national power, particularly diplomacy, economic power, and even information type approaches, are effective tools for maneuvering through these tricky waters.

Audience: Earlier, after speculating about forming an oil embargo against China, you said that you didn't think it was possible. How secure is the energy that is being transported around the world? One could envisage a terrorist attack or a small ad hoc attack on a facility, a pipeline, or a choke point that would cause a short-term disruption. But aside from that, is there any force or group of countries capable of producing a long-term disruptive effect on oil successfully that couldn't be countered by other interested parties?

Audience: The only possible source of a massive disruption of seaborne supplies of energy is the U.S. Navy, and I don't think that's going to happen.

CONCLUSIONS

Douglas Lovelace: First of all, I thought the conference provided a lot of technical information for those of us who work in national security strategy and national security policy, providing a good foundation for understanding numerous energy issues. It was very valuable from that perspective.

I'm not sure whether we really looked at the nexus of energy and security as much as we looked at energy security itself. This may be a subtle distinction, but there is a difference between these topics. However, I think that we will probably see that nexus delineated more clearly in the published proceedings thanks to comments that were made during the course of the conference.

William Boettcher: As we were coming up with ideas for the conference, we struggled—Vik Rao, Chris Gould, Man-Sung Yim, Carolyn Pumphrey, and I—over what we thought energy security meant and what we thought the energy and security nexus was. To me, it's the multidimensional aspect of this issue that's been highlighted by the conference. We, of course, forced that to happen because we invited all of you to provide different perspectives.

At one level we see Eugene Gholz's definition of energy security, which more or less equates to prosperity for the United States, and which involves these notions of stability and availability of energy at reasonable prices. But there are other aspects to energy security. For one thing, there is the degree to which energy production and consumption affect physical security. We have talked some about the number of troops, the number of Soldiers, Sailors, Marines, and Airmen that are deployed around the world to secure our energy supplies or protect those energy supplies.

Energy security also affects environmental security, and we talked a little bit about its relationship to climate change. Alex Roland asked the question, Why now? Why are we talking about energy security today? (Remember, we were planning the conference long before the recent events in the Middle East and rise in oil prices.) The answer in part is that there now seem to be alternatives to a fossil fuel-based economy. Has that solved some of the security dilemmas we face? Has it made the Middle East less important? Will increased energy production improve the U.S. domestic economy? Will it lead to less CO_2 production? Could it affect climate change which in turn would have a security impact?

We are also interested in the human security dimension, which means considering security beyond the United States. How do all of these decisions — our consumer choices; government regulation; behavior by us, by big oil, and by the governments of other countries — affect human beings around the world in terms of food security, economic security, and some of these other dimensions?

The point that struck me is that there are going to be trade-offs. It is unlikely that we're going to find any solution that's zero cost or low cost. We're not going to find an alternative energy source that approximates the quality or performance characteristics of what we now have. Therefore, if we are going to move away any time soon from this fossil fuel economy, we are going to have to balance the resulting good and the bad. There are going to be security, economic, and political tradeoffs.

The final point I want to make is about so-called political reality. I've heard a lot about political reality, and I just don't believe it exists. We ourselves construct that reality: politicians do it, elites do it, the kind of people in this room do it. You can transform that reality. George Bush, who said the American lifestyle is not negotiable, also said, "Read my lips, I'm not going to raise taxes," and then he did. When we're talking about human behavior and human agency, you can promote change. I remember 2 years ago, there was this new reality when Democrats enjoyed a landslide election and had a mandate. Obama was going to change the world. So I am not as pessimistic as some about changing some of those behaviors.

Vikram Rao: I think of security largely as a matter of reducing imported oil and, in so doing, creating jobs at home. It'll also benefit the country by mending our balance of payments problem. If we reduce imported oil, more than likely it will mean we'll get less of it from far-off places. We'll probably get more of it from Canada and Mexico. But either way, if you replace imported oil with some other form of fuel, jobs are created here, and that's what I seek.

The other point I would make is that economic security equates to other forms of human security. One of the significant events in this nation is the discovery of shale gas and the possibilities this gives us of becoming more gas self-sufficient. Whether we become exporters or not is just a detail. The implications of these shale gas finds are huge.

When we had the gas price spike a few years ago, industry actually left the country—polypropylene manufacturers went abroad. That's because natural gas is the basic building block of more things than you would care to believe. Coal can be too, but natural gas already is. Essentially most of your fertilizers, almost every item of clothing you wear other than cotton, and every package you have comes from natural gas. All the methanol proposed as an alternative auto fuel comes from natural gas. Whatever depletes or increases natural gas in this country will have profound economic security implications. For example, if natural gas was plentiful and cheap, methanol, or diesel fuel, or even jet fuel made with alternatives becomes more viable. It would not be viable with liquefied natural gas (LNG) imports. LNG has a floor price of $4 a million BTUs, just because that's what it costs to compress, ship, and decompress it.

In sum, energy security means two things: reducing imported oil by replacing it with substitutes, whether it's electricity or whatever; and using the newfound natural gas effectively. Obviously, natural gas presents environmental issues of its own, and we have to deal with them.

ENDNOTES - CHAPTER 6

1. A compact fluorescent (CF) lamp is a fluorescent lamp designed to use less power (typically one-fifth) and has a longer rated life (six to 10 times average) than traditional incandescent lamps. However, they contain mercury, which makes disposal a problem.

2. According to the International Energy Agency (IEA) "Energy Security can be described as "the uninterrupted physical availability at a price which is affordable, while respecting environment concerns."

ABOUT THE CONTRIBUTORS

JAMES T. BARTIS is a senior policy researcher at the RAND Corporation. He joined the U.S. Department of Energy (DoE) in 1978 shortly after it was established. He served in the Office of Fossil Energy, and in DoE's main policy office. During the George H. W. Bush and Clinton administrations, he was a member of the Industry Sector Advisory Committee on Energy for Trade Policy Matters. Dr. Bartis has more than 25 years of experience in policy analyses and technical assessments in energy and national security. His recent energy research topics include analyses of the international petroleum supply chain, assessments of alternative fuels for military and civilian applications, development prospects for coal-to-liquids and oil shale, energy and national security, Qatar's natural gas-to-diesel plants, Japan's energy policies, planning methods for long-range energy research and development, critical mining technologies, and national response options during international energy emergencies. Dr. Bartis has written some thirty monographs, most recently *Alternative Fuels for Military Applications* (2011); *Managing Spent Nuclear Fuel: Strategy Alternatives and Policy Implications* (2010); and *Near-Term Feasibility of Alternative Jet Fuels* (2009). Dr. Bartis holds a Ph.D. in chemical physics from the Massachusetts Institute of Technology.

WILLIAM A. BOETTCHER III is an Associate Professor of Political Science and Public Administration at North Carolina State University (NCSU) and NCSU representative to the Executive Board of the Triangle Institute for Security Studies (TISS). His research focuses on the management of risk in foreign policy decisionmaking and the framing of casualty data. Dr. Boettcher is the author of several works, including *Presidential Risk Behavior in Foreign Policy: Prudence or Peril* (2005). He is a key player in the TISS/NCSU Energy and Security Initiative and is currently working with the Nuclear Engineering Department to build a program in nuclear security. He team-taught a pilot course with Man-Sung Yim in the Spring 2010 and has written and taught about weapons of mass destruction. Dr. Boettcher holds a Ph.D. in political science from Ohio State University.

KEVIN BOOK is Managing Director of Research at Clearview Energy Partners, LLC. The group provides corporate and financial-sector clients with analyses of how government actions affect energy fundamentals and offers guidance to help financial and physical players in the sector minimize risk and maximize value. His primary coverage areas include oil and natural gas refining; climate policy; alternative fuels, vehicles, and power; and geopolitical risk analysis. Mr. Book testified before the U.S. Senate Committee on Energy and Natural Resources in May 2009 and before the U.S. Senate Committee on Finance's subcommittee on Energy, Natural Resources, and Infrastructure in May 2010. Prior to co-founding ClearView in February 2009, he worked as Senior Vice President of energy policy, oil, and alternative energy research at FBR Capital Markets Corporation. Mr. Book holds an M.A. in law and diplomacy from the Fletcher School of Law and Diplomacy, and a B.A. in economics from Tufts University.

JOHN BUMGARNER is Chief Technology Officer at the U.S. Cyber Consequences Unit (USCCU) an independent, non-profit research organization that investigates the strategic and economic consequences of cyber attacks. He has over 20 years of work experience in information security, military special operations, intelligence, and physical security. In his work as a security researcher, Mr. Bumgarner has discovered cyber vulnerabilities, uncovered previously undetected cyber attacks, and invented new cyber-attack techniques. He has served as an expert source for national and international news organizations, and his articles have appeared in the journal of the *Information System Security Association*, the *Homeland Security Journal*, the *Information Operations Journal*, the *Counter Terrorist* magazine, and other leading security publications. Mr. Bumgarner is the author of the U.S. Cyber Consequences Unit's analysis of the August 2008 cyber campaign against Georgia, and the co-author, along with Scott Borg, of the USCCU Cyber Security Check List, which is currently used by cyber security professions in over 80 countries. Mr. Bumgarner holds private sector certifications including Certified Information Systems Security Professional (CISSP), Global Information Assurance Certification (GIAC)—Gold level, and dual master's degrees in information systems management and security management.

ROBERT F. CEKUTA is the Principal Deputy Assistant Secretary of the Energy Resources Bureau, U.S. Department of State. Prior to this, he was the Deputy Assistant Secretary, Energy, Sanctions, and Commodities, U.S. Department of State. In this role, he was responsible for energy security and other aspects of U.S. global energy policy as well as for sanctions regimes and combating terrorists' efforts to misuse the international financial system. He was also responsible for matters affecting the availability of key commodities. Recent overseas assignments include Tokyo (2007-09) and Berlin (2003-07) where he led the U.S. Government's engagement on the full range of economic issues with two of the world's top economies. Mr. Cekuta's work as Economic Minister Counselor in Germany included counterterrorism and efforts to combat international criminal activities. He was also Senior Advisor for Food Security in the State Department's Bureau of Economic, Energy, and Business Affairs and Senior Deputy Coordinating Director at the U.S. Embassy in Kabul, Afghanistan for all development and economic affairs. Having been assigned overseas in Albania, Austria, Iraq, South Africa, and Yemen, Mr. Cekuta has also worked in the Office of the U.S. Trade Representative (1999-2000). He has held a variety of other positions in Washington at the State Department responsible for energy, trade, and economic development. Mr. Cekuta holds master's degrees from the Thunderbird School of Global Management and the National War College.

BERNARD D. COLE is Professor of International History at the National War College in Washington, DC, where he concentrates on the Chinese military and Asian energy issues. Dr. Cole has written numerous articles and six books, several of which focus on energy issues. He is the author of *Sea Lanes and Pipelines: Energy Security in Asia* (2008). His 2003 book, *Oil for the Lamps of China: Beijing's 21st Century Search for Energy* (2003), examined China's dependence on foreign sources of energy supplies and the resulting national security implications. His latest work, *The Great Wall at Sea: China's Navy in the Twenty-First Century* (2010), focuses on China's maritime tradition, and discusses the nation's increasing dependence on energy sources mined from the ocean floor. A retired Captain of the U.S. Navy, Bernard Cole received a Ph.D. in history from Auburn University.

DAVID C. DAYTON is the director of chemistry and Biomass Program manager in the Research Triangle Institute's (RTI) Center for Energy Technology. He has over 15 years of project management and research experience in biomass thermo-chemical conversion research and development. Prior to joining RTI, he was the thermo-chemical platform leader at the National Renewable Energy Laboratory. Dr. Dayton has extensive experience investigating fundamental, high-temperature kinetics of thermo-chemical conversion processes. He also has extensive research and development experience related to cleanup and conditioning of biomass-derived synthesis gas, specifically catalytic steam reforming of tars, to provide a suitable feedstock for the production of renewable fuels and chemicals. Dr. Dayton came to RTI in 2007 following 14 years at the National Renewable Energy Laboratory. He was a postdoctoral research associate at the U.S. Army Research Laboratory, Aberdeen Proving Ground, Maryland, from 1991 to 1993. Dr. Dayton holds a B.S. in chemistry from Dickinson College, and a Ph.D. in physical chemistry from the University of North Carolina, Chapel Hill.

EUGENE GHOLZ is an Associate Professor of Public Affairs at the Lyndon B. Johnson School of Public Affairs at the University of Texas (UT) at Austin, and Associate Director for Policy at UT's new Center for Energy Security. He is a Research Associate at the Massachusetts Institute of Technology's (MIT) Security Studies Program, a life member of the Council on Foreign Relations, and associate editor of the journal *Security Studies*. He is currently working in the Office of Industrial Policy at the U.S. Department of Defense. Dr. Gholz has written on innovation, business-government relations, defense management, U.S. foreign military policy, and energy security issues. He is the author and co-author of many works, including *Buying Transformation: Military Innovation and the Defense Industry* (2006) and *US Defense Politics: The Origins of Security Policy* (2009). Dr. Gholz holds a Ph.D. in political science from the Massachusetts Institute of Technology.

CHRISTOPHER GOULD is Associate Dean for Administration at the College of Physical and Mathematical Sciences at North Carolina State University. He is an Alumni Distinguished Undergraduate Professor of Physics and previously served for a decade as Head of the Department of Physics at NCSU. He is a nuclear physicist by training, with interests in cosmology, energy research and policy, science education, and neutron and neutrino physics. Dr. Gould has held visiting appointments at Los Alamos National Laboratory; the Institut für Kernphysik, Frankfurt; the Atomic Energy Institute, Beijing; the University of Petroleum and Minerals, Dhahran; and the Oak Ridge Center for Advanced Studies. He is a Fellow of the American Physical Society and a member of the American Association of Physics Teachers and of Sigma Xi. Dr. Gould has a bachelor's degree from Imperial College, London, and a Ph.D. from the University of Pennsylvania.

ALAN S. HEGBURG is a senior fellow in the Center for Strategic and International Studies (CSIS) Energy and National Security Program. Prior to joining CSIS, he served for 17 years in the U.S. Departments of State and Energy, most recently as deputy assistant secretary for international energy policy in the Office of Policy and International Affairs at the Energy Department. In that position, he directed analytical, energy policy, and representational work covering the Eurasian land mass (including Russia, the Baltics, and Ukraine), Central Asia, the Caspian and the Caucasus, the Middle East, Africa, and certain multilateral organizations. At the Department of State, Mr. Hegburg served in the Europe and Asia Bureaus and internationally in Germany and Paris, the latter on assignment to the International Energy Agency. In the private sector, he spent 17 years at Phillips Petroleum, Amoco, and British Petroleum as manager of international government relations and director of international and geopolitical analysis. He has taught graduate courses on the geopolitics of energy at George Washington and Columbia Universities. Mr. Hegburg holds a master's degree from the Johns Hopkins University.

RICHARD C. KEARNEY is Director of the School of Public and International Affairs at North Carolina State University. His research and teaching fields are in public administration, human resource management, labor relations, state and local politics, and public policy (particularly environmental policy). He has spoken and written on the challenges posed by nuclear waste. Dr. Kearney's recent published work includes *State and Local Government* (8th Ed.); *Public Human Resource Management: Problems and Prospects* (6th Ed.); *Labor Relations in the Public Sector* (4th Ed.); and articles in *Journal of Labor Research, Publius, Review of Public Personnel Administration,* and *Public Administration Review.* Dr. Kearney has previously served at East Carolina University, the University of Connecticut, and the University of South Carolina. He has taught and conducted research for extended periods in the Dominican Republic and Mexico, and was a Fulbright senior lecturer at the University of Mauritius. Dr. Kearney holds a Ph.D. in political science from the University of Oklahoma.

ROSEMARY A. KELANIC is Associate Director of the Institute for Security and Conflict Studies, Elliott School of International Affairs, George Washington University. Prior to this, she was an International Security Program Research Fellow at the Belfer Center for Science and International Affairs, Harvard University. Her research interests include international relations theory, energy security, and resource conflict. As a Grodzins Prize Lecturer at Chicago, she recently designed and taught an advanced undergraduate course on oil and international security. Dr. Kelanic holds a B.A. in political science from Bryn Mawr College, and an M.A. in international relations from the University of Chicago. She holds a Ph.D. in political science from the University of Chicago.

STEPHEN R. KELLY is a Visiting Professor at the Sanford School of Public Policy and Center for Canadian Studies at Duke University. His specialty areas are energy, security, and North American issues, including trade, immigration and border management, and he is currently teaching a course at the Sanford School of Public Policy on Energy Security. Mr. Kelly came to Duke as Diplomat in Residence at the end of 28-year career in the Foreign Service. Prior to this assignment, he was Director of the Senior Level Assignments Division at the State Department in Washington, DC. From 2004 to 2006, he was Deputy Chief of the U.S. Mission to Mexico, and from 2000 to 2004, Mr. Kelly was Deputy Chief of the U.S. Mission to Canada. He also served as Consul General in Quebec City from 1995 to 1998, where he was the chief U.S. reporting officer on the Quebec Sovereignty Referendum of October 1995. Other overseas postings include the Netherlands as political counselor, Indonesia as human rights officer, Belgium as a political and consular officer, and Mali, West Africa, as a management officer. Mr. Kelly is a graduate of Cornell University and holds an M.A. in national security strategy from the National War College.

CAREY KING is Research Associate at the Center for International Energy and Environmental Policy, the Jackson School of Geosciences at the University of Texas at Austin. He studies energy systems, how they work together, how they impact the environment, and how and why humans as a society consume energy resources. He specializes in dynamic systems modeling and nonlinear optimization methods. Much of Dr. King's recent work has focused on the nexus between energy and water for projecting water demand for electricity generation and alternative automobile fuels. Additional work in which he is engaged includes the economics of carbon capture and sequestration, the design of beneficial combinations of renewable energy and storage systems, and the creation of tools to help the public and policymakers understand the tradeoffs among different electricity generation sources. Dr. King is a former National Science Foundation Graduate Fellow, and holds both a B.S. and Ph.D. from the Department of Mechanical Engineering at the University of Texas at Austin.

ANNE KORIN is co-director of the Institute for the Analysis of Global Security (IAGS). She chairs the Set America Free Coalition, a strange bedfellow alliance spanning the political spectrum focused on policies to reduce the strategic importance of oil. She also chairs the Mobility Choice for a Secure America Coalition, an alliance promoting a fiscally responsible, free market oriented approach to expanding competition among transportation modes. Ms. Korin appears in the media frequently and has written articles for *Foreign Affairs, The American Interest, MIT Innovations, American Legion Magazine, The National Review, Commentary Magazine,* and the *Journal of International Security Affairs.* She is co-author of *Energy Security Challenges for the 21st Century* (2009) and *Turning Oil into Salt: Energy Independence through Fuel Choice* (2009). Ms. Korin holds an engineering degree in computer science from Johns Hopkins University, and is working toward a Ph.D. from Stanford University.

DOUGLAS C. LOVELACE, JR., is the Director of the Strategic Studies Institute at the U.S. Army War College. His Army career included a combat tour in Vietnam and a number of command and staff assignments. While serving in the Plans, Concepts, and Assessments Division and the Conventional War Plans Division of the Joint Staff, he collaborated in the development of documents such as the National Military Strategy, the Joint Strategic Capabilities Plan, the Joint Military Net Assessment, national security directives, and presidential decision directives. He has published extensively in the areas of national security and military strategy formulation, future military requirements and strategic planning. Professor Lovelace holds a J.D. from Widener University.

STEVEN E. MILLER is the Director of the International Security Program, Editor-in-Chief of the quarterly journal, *International Security*, and also co-editor of the International Security Program's book series, Belfer Center Studies in International Security, Harvard. Previously, he was Senior Research Fellow at the Stockholm International Peace Research Institute (SIPRI) and taught Defense and Arms Control Studies in the Department of Political Science at the Massachusetts Institute of Technology. Dr. Miller currently co-directs the Academy's project "On the Global Nuclear Future." In this capacity, he has co-chaired two conferences on the regional implications of the nuclear renaissance, one in Abu Dhabi (November 2009) and the other in Singapore (November 2010). Dr. Miller is co-author of a work on the war with Iraq and editor or co-editor of more than two dozen books, including, most recently, *Going Nuclear* (January 2010) and *Contending with Terrorism* (July 2010). Dr. Miller holds a M.A. in law and diplomacy and a Ph.D. in international relations from the Fletcher School of Law and Diplomacy.

CAROLYN PUMPHREY has served as Program Coordinator for the Triangle Institute for Security Studies (TISS) since 2000 and also teaches history at North Carolina State University (NCSU). Over the last 13 years, she has organized over 30 conferences and workshops for TISS. From 1997-2000, she was a Post-Doctoral Fellow for the TISS and, between 1986 and 1992, she was an Assistant Professor of History at Spring Hill College in Mobile, Alabama. Dr. Pumphrey has taught a wide range of courses, including the "History of Restraints on War," "War and Peace in Ancient and Medieval Times," "Medieval Warfare," and "In War and Peace: Christian and Muslim Relations in the Middle Ages." She has edited four books, *Transnational Threats: Blending Law Enforcement and Military Strategies* (2000); *The Rise of China: Security Implications* (2002); (with Rye Schwartz-Barcott), *Armed Conflict in Africa* (2003); and *Global Climate Change: National Security Implications* (2008). Dr. Pumphrey holds a B.A. in literature and history from the University of York, England, and a Ph.D. in history from Duke University, where she was the Medieval and Renaissance Studies Fellow (1978-1981).

VIKRAM RAO is the Executive Director of the Research Triangle Energy Consortium (RTEC). RTEC was formed in 2007 by the University of North Carolina at Chapel Hill, Duke University, North Carolina State University, and the Research Triangle Institute (RTI) International to highlight energy issues, better inform academics, industry, policymakers and investors, and conduct research and development to create sustainable sources of energy. Dr. Rao is the author of more than 40 publications and has been awarded 24 patents in fields that include nonferrous metal refining, alloy formulations, and oil and gas technology. He serves in technical and business advisory capacities for energy companies, technology companies, and nongovernmental organizations and universities in the United States and elsewhere. He is the producer/writer of the Director's Blog. Dr. Rao holds a bachelor's degree in engineering from the Indian Institute of Technology in Madras, India, and a master's and Ph.D. in engineering from Stanford University.

MICHAEL J. ROBERTS is Associate Professor in the Department of Agricultural and Resource Economics, North Carolina State University. He is also an Affiliate of the Center for Environmental and Resource Economic Policy. He has served as the Director of the Northeastern Agricultural and Resource Economics Association and has been an Adjunct Professor at the Paul H. Nitze School of Advanced International Studies, Johns Hopkins University (2005-07) and an Instructor at the City College of San Francisco (1997). Dr. Roberts has written and spoken on the impact of climate change on crops and economies, on energy prices in general, and on the biofuel mandate and world food prices in particular. Dr. Roberts holds a B.A. in quantitative economics and decisions sciences from the University of California at San Diego, an M.A. in statistics from the University of California at Berkeley, an M.S. in applied economics from Montana State University, and a Ph.D. in agricultural and resource economics from the University of California at Berkeley.

ALEX ROLAND is a Professor Emeritus of History at Duke University. Between 1973 and 1981, he was a historian at the National Aeronautics and Space Administration. From 1988 to 1989, he was the Harold K. Johnson Visiting Professor of Military History at the Military History Institute, U.S. Army War College. From 2001 to 2002, he was the Dr. Leo Shifrin Professor of Naval-Military History at the U.S. Naval Academy. He is a retired Marine Corps officer. Dr. Roland's research and writing have been in the fields of aviation, astronautics, computers, weapons, and the relationship between war and technology. Among his publications are *Strategic Computing: DARPA and the Quest for Machine Intelligence*, 1983-1993 (2002), *The Military Industrial Complex* (2001), *Atmospheric Flight in the Twentieth Century* (edited with Peter Galison, 2001), and most recently (with W. Jeffrey Bolster and Alexander Keyssar), *The Way of the Ship* (2008). He is a past President of the Society for the History of Technology. Dr. Roland holds a Ph.D. in history from Duke University.

JAMES A. TRAINHAM is the Vice President of Strategic Energy initiatives at Research Triangle International. He also holds a joint appointment at North Carolina State University. Dr. Trainham's focus is on the development of solar fuels as part of a new solar fuels institute in the Research Triangle Park. He specializes in both product and process technologies, including alternative energy, specialty chemicals, coatings, polymers, and synthetic fibers. Before RTI, Dr. Trainham directed research and development, engineering design, and scale-up for Sundrop Fuels, Inc., as senior vice president. He served as vice president of Science and Technology at PPG Industries for 4 years and had a 25-year career at the DuPont Company. He has been elected to the National Academy of Engineering. Dr. Trainham holds a Ph.D. in chemical engineering from the University of California-Berkeley.

DANIEL J. WEISS is a Senior Fellow and the Director of Climate Strategy at the Center for American Progress (CAP), where he leads the Center's clean energy and climate advocacy efforts. Among issues he has addressed is the need for the U.S. Air Force to develop a clean energy strategy. Before coming to CAP, he spent 25 years working with environmental advocacy organizations and political campaigns. Most recently, he was a senior vice president with M+R Strategic Services, where he oversaw collaborative campaign efforts by 15 major national environmental organizations working to oppose anti-environmental policies. Prior to M+R Strategic Services, Mr. Weiss served for 16 years at the Sierra Club, first as a Washington representative, then as director of the Environmental Quality Program, and for the final 8 years as political director. Mr. Weiss holds an M.P.P. degree from the University of Michigan.

MAN-SUNG YIM is Professor and Department Head, Nuclear and Quantum Engineering at the Korean Advanced Institute of Science and Technology (KAIST). Prior to assuming this position, he was Associate Professor of Nuclear Engineering and member of the Associated Faculty of Civil, Construction & Environmental Engineering at North Carolina State University. He is interested in guiding and supporting the development of back-end nuclear fuel cycle technologies and effective nuclear nonproliferation regimes. The goal of his research work is to minimize human health risk, environmental impact, and nuclear proliferation risk from the use of civilian nuclear power technology. His research uses mathematical/statistical modeling, engineering optimization, data mining, risk analysis, decision theoretic approaches, and policy analysis. Dr. Man-Sung Yim has written and taught widely on the nuclear fuel cycle, nuclear waste management, environmental exposure and risk, and nuclear nonproliferation policy. Dr. Man-Sung Yim holds an S.M. in environmental health science, a Sc.D. in radiological health from Harvard, undergraduate degrees in engineering from Seoul National University, and a Ph.D. in nuclear engineering from the University of Cincinnati.

www.ingramcontent.com/pod-product-compliance
Lightning Source LLC
Chambersburg PA
CBHW071358170526
45165CB00001B/95